THE FOURIER INTEGRAL
AND CERTAIN OF ITS
APPLICATIONS

THE FOURIER INTEGRAL AND CERTAIN OF ITS APPLICATIONS

BY

NORBERT WIENER
Professor of Mathematics at the Massachusetts Institute of Technology

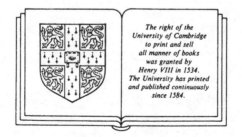

CAMBRIDGE UNIVERSITY PRESS
Cambridge
New York   New Rochelle
Melbourne   Sydney

CAMBRIDGE UNIVERSITY PRESS
Cambridge, New York, Melbourne, Madrid, Cape Town, Singapore, São Paulo, Delhi

Cambridge University Press
The Edinburgh Building, Cambridge CB2 8RU, UK

Published in the United States of America by Cambridge University Press, New York

www.cambridge.org
Information on this title: www.cambridge.org/9780521358842

© Cambridge University Press 1933
Reissue with Foreword © Cambridge University Press 1988

This publication is in copyright. Subject to statutory exception
and to the provisions of relevant collective licensing agreements,
no reproduction of any part may take place without the written
permission of Cambridge University Press.

First published 1933
Reissued with Foreword in the Cambridge Mathematical Library series 1988
Re-issued in this digitally printed version 2008

A catalogue record for this publication is available from the British Library

ISBN 978-0-521-35884-2 paperback

DEDICATED
TO THE MEMORY OF
CLARENCE LEMUEL ELISHA MOORE
*Late Professor of Mathematics at the Massachusetts
Institute of Technology*

CONTENTS

Foreword — page ix
Preface — xv

INTRODUCTION

§ 1 The Nature of Harmonic Analysis — 1
2 The Properties of the Lebesgue Integral — 4
3 The Riesz-Fischer Theorem — 27
4 Developments in Orthogonal Functions — 34

CHAPTER I
PLANCHEREL'S THEOREM

§ 5 The Formal Theory of the Fourier Transform — 46
6 Hermite Polynomials and Hermite Functions — 51
7 The Generating Function of the Hermite Functions — 55
8 The Closure of the Hermite Functions — 64
9 The Fourier Transform — 67

CHAPTER II
THE GENERAL TAUBERIAN THEOREM

§ 10 Enunciation of the General Tauberian Theorem — 72
11 Lemmas Concerning Functions whose Fourier Transforms Vanish for Large Arguments — 80
12 Lemmas on Absolutely Convergent Fourier Series — 86
13 The Proof of the General Tauberian Theorem — 94
14 The Closure of the Translations of a Function of L_1 — 97
15 The Closure of the Translations of a Function of L_2 — 100

CHAPTER III

SPECIAL TAUBERIAN THEOREMS

§ 16 The Abel-Tauber Theorem	*page* 104

17 The Prime-Number Theorem as a Tauberian Theorem	112

18 The Lambert-Tauber Theorem	119

19 Ikehara's Theorem	125

20 The Mean Square Modulus of a Function	138

CHAPTER IV

GENERALIZED HARMONIC ANALYSIS

§ 21 The Spectrum of a Function	151

22 The Spectra of Certain Linear Transforms of a Function	164

23 The Monotoneness of the Spectrum	180

24 The Elementary Properties of Almost Periodic Functions	185

25 The Weierstrass and Parseval Theorems for Almost Periodic Functions	196

Bibliography	200

FOREWORD

This is a most inspired and inspiring book.

When it was written two of Wiener's major papers had just appeared: Generalized Harmonic Analysis, and Tauberian Theorems. His previous works on potential theory, Brownian motion, Fourier analysis were highly appreciated by a few dozens of mathematicians in the world. It was a happy time in the life of Norbert Wiener. He was thirty-seven years old, he had been married for six years, he had just been promoted to a professorship at the Massachusetts Institute of Technology, and he had spent a pleasant year in Cambridge, England, with his wife Margaret and their small daughters, Barbara and Peggy.

Cambridge (I mean, Cambridge, England) plays a special role in Wiener's life. The beginning of his mathematical career coincided with his meeting with Bertrand Russell in Cambridge in 1913. He had just graduated from Harvard in mathematical logic – at only eighteen – and was supposed to go on in logic with Russell. Actually Russell's advice was to learn more mathematics and physics. Accordingly, Norbert Wiener took courses with Hardy, Littlewood, Mercer, and read Einstein's papers of 1905, and Niels Bohr's recent works. Without any doubt Einstein's theory of Brownian motion had a decisive role in Wiener's inspiration.

The academic year 1931–32 in Cambridge was an opportunity for Wiener to meet Hardy and Littlewood again, to lecture on Fourier integrals (in Hardy's chair) and to discover an ideal collaborator, the young R. E. A. C. Paley. Paley had already worked with Littlewood and with Zygmund, and was to die, one year later, in a skiing accident when visiting Wiener at MIT. Two books resulted from the stay in Cambridge and the collaboration with Paley: the present book, which deals with real variable aspects of the Fourier transform, and its superb companion, *Fourier Transforms in the Complex Domain*, written by Wiener after Paley's death, with both as coauthors.

FOREWORD

These two books contain the essential ideas of Wiener in harmonic analysis and should be read by anyone working in the field. The second one also has two chapters on random functions with the most important results of Wiener on Brownian motion. Wiener was still to work and discover new things in harmonic analysis (in particular, on Fourier–Stieltjes transforms). However, his main interests shifted gradually to prediction theory, a topic at the frontier of harmonic analysis, and to cybernetics. Out of the circle of mathematicians Norbert Wiener is known as the founder of cybernetics. However, for mathematicians, his most prominent work is in and around harmonic analysis.

Harmonic analysis as we see it now is essentially harmonic analysis as it was viewed by Wiener. Translations of functions (the term of translation, applied to a function, was introduced by Wiener) and convolutions (not yet named in English, for Wiener uses the German word *Faltung*) play a basic role. From the point of view of physicists and engineers, convolutions appear whenever a signal is transformed into an observation through an apparatus: the convolution kernel defines the action of the apparatus. From a mathematician's viewpoint regularization of functions and summation processes for trigonometric series or integrals are nothing but convolutions.

Convolutions appear also, in a hidden form, when some summation processes are applied to ordinary series, integrals, or mean values. Wiener had examples coming from his theory of generalized harmonic analysis. His friend A. E. Ingham (another connection with Cambridge!) led him to a beautiful theorem of Littlewood (generalizing a rather simple result of Tauber) which provides another example. Finally, the prime number theorem can be approached in the same manner. The unifying tool is what Wiener called, as a homage to Littlewood, the general Tauberian theorem.

The Tauberian theorem states that when a limit exists in a certain way, it exists also in another way. What defines the way is a convolution kernel. In a more abstract form, the Tauberian theorem says that, under some conditions, the second kernel belongs to the span of the translates of the first (what Wiener

FOREWORD

calls its 'extension'). Now comes the Fourier transform. If the Fourier transform of the first kernel does not vanish, its extension contains all kernels. In other words, $L^1(\mathbb{R})$ is generated by the translates of a function whenever its Fourier transform does not vanish. To prove this, a detour through Fourier series (looking at $L^1(\mathbb{Z})$ instead of $L^1(\mathbb{R})$) proves convenient: the Fourier transforms become sums of absolutely convergent Fourier series (the class A of Wiener) and the basic lemma says that, whenever a function belongs to A and does not vanish, its inverse belongs to A.

Not only did the general Tauberian theorem give a unifying view on questions involving summations and limits, but it introduced a paradigm for what was called abstract harmonic analysis a few years later. The class A is a normed ring in the sense of Gelfand, a Banach algebra as we say now. Analytic functions operate on A (this is the Wiener–Lévy theorem) and questions about endomorphisms of A, operating functions, structure of closed ideals, became famous in the years 1940–60. Wiener converted absolutely convergent Fourier series into a central subject in analysis.

The Tauberian theorems occupy a central place in the present book. The last application contains the first motivation, a formula suggested by generalized harmonic analysis. Generalized harmonic analysis is the subject-matter of the last chapter, though it was conceived before the Tauberian theorems. Roughly speaking, it consists in studying a large class of functions for which a 'spectrum' can be defined. Already Bohr's and Besicovitch's theories of almost periodic functions had given a good mathematical model for natural phenomena with discrete spectra. However, Wiener argues that most spectra occurring in physics are continuous, or contain both a discrete and a continuous part. Wiener's model consists of functions $f(x)$ for which the mean value of $f(x)\overline{f(x+\xi)}$ exists whatever the translation ξ may be. This mean value, $\phi(\xi)$, divided by $\phi(0)$, expressed a kind of correlation between the function and its translate by ξ. Bochner's theory of positive definite functions applies to ϕ and proves that ϕ, when it is continuous, is the Fourier transform of a positive measure, exactly the spectrum of Wiener. Actually Bochner's theory was not available to Wiener, and he had to build up his own method in order to exhibit the spectrum. The example he gives of a function with a continuous spectrum is a random function – one of the first examples of probability methods applied to a problem in analysis. The end of the chapter deals with Bohr's almost periodic functions, proving in particular the Plancherel formula for these functions. Plancherel's formula expresses an L^2-isometry which is like Ariadne's thread, running through Wiener's mathematical world.

Plancherel's formula in the context of $L^2(\mathbb{R})$ is the subject of the first chapter of the book. The original treatment of Wiener through Hermite functions was later the basis for the theory of homogeneous chaos: here again we see the connection between his views on stochastic processes and questions in harmonic analysis. Actually the so-called Wiener integral, like the Fourier transform, has the fundamental property of defining an isometry in L^2-spaces. A modern introduction of Brownian motion can start from this fact.

When Wiener's book appeared, Fourier series and integrals were a rather popular subject in mathematics. Several books by first class mathematicians contributed to the field. Wiener's book has a special flavour. It is not really a treatise on the Fourier integral. It is not intended to pack and dispatch as much material as possible. It does not look for the shortest ways to prove theorems. It is highly original, with emphasis on a selection of important topics, with unifying views and methods, large perspective and careful attention to details. It is a most attractive expression of his personality.

This is the third printing of the book. Wiener would have welcomed it. He was very sensitive to the reactions of mathematicians towards his papers and he resented it whenever there was a lack of recognition of his work. It might have been the case with the reviews on this book. Surprisingly, it got only eighteen lines of analysis and comment in *Zentralblatt für Mathematik* when it appeared, and two lines in *Mathematical*

FOREWORD

Reviews for the second printing, while very poor papers are reviewed at great length. No matter what kind of reviews it got, this little book is one of the great mathematical achievements of the century.

November, 1987   Jean-Pierre Kahane

PREFACE

The present book is in substance an elaboration of a course of fifteen lectures on the Fourier Integral and its Applications, given at the University of Cambridge during the Lent Term of 1932. When I arrived in Cambridge during the Michaelmas Term of 1931, on leave of absence from the Massachusetts Institute of Technology, I had vague plans of writing up certain topics in the theory of harmonic analysis into a book on the subject. My original idea was of a rather comprehensive treatise, proceeding from the elements of Lebesgue integration through the L_2 theory of the Fourier series to the Plancherel theorem, the Fourier Integral, the periodogram, and lastly, to theorems of Tauberian type. My impulse to write a book of this type arose from a dissatisfaction with the preponderant rôle of convergence theory in existing textbooks on the subject, and from the need for a treatment more in line with the extensive periodical literature.

As far as my desire to write a book sprang from the need for a textbook to use in my course at the Massachusetts Institute of Technology, it has largely been dissipated by the recent appearance of a book on the Theory of Functions by Professor Titchmarsh. Several chapters of his book are devoted to the treatment of Fourier series from the modern point of view. Unfortunately—from my standpoint—he does not allot a great deal of space to the Fourier Integral and related matters. Thus, while there is now no need for the comprehensive treatise which I at first contemplated, there is need for a discussion of the Fourier Integral from the modern point of view. When Professor Titchmarsh's book has been in use for some five years, and has become the basis for higher instruction in Fourier series, it will be possible to treat the Fourier Integral in a thoroughgoing and coordinate way, but for the present we shall have to content ourselves with more fragmentary treatments.

Thus when Mr Besicovitch and Professor Hardy suggested my giving a course during the Lent Term, I gladly fell in with their plans, and offered as a topic the Fourier Integral and its Applications. It was none of my purpose to aim at completeness, but merely to present various aspects of the theory whose sole unity was that I had worked in all of them. When Professor Hardy later suggested that I should submit the manuscript to the Cambridge University Press, we both agreed that at this time it was better to make the book a frank course of lectures than to strive for the clean-cut outline of a treatise. There are three more or less separate groups of ideas which accordingly find representation: the group pertaining to the Fourier transform and the Plancherel theorem; the notions of an absolutely convergent Fourier series and of a Tauberian theorem; and the concept of the spectrum. This last idea is, it is true, dependent on both the preceding parts of the book, and serves to give it some degree of unity.

The main results of the book may be listed as follows:

(1) The Plancherel theory of the existence of the Fourier transform of a function of L_2, together with the associated Parseval theorem and the proof of the theorem yielding the inverse of a Fourier transformation;

(2) The theorem asserting that if $f(x)$ is a continuous non-vanishing function with an absolutely convergent Fourier series, the Fourier series of $1/f(x)$ converges absolutely;

(3) Various forms of general Tauberian theorems;

(4) The Lambert-Tauber theorem and the de la Vallée Poussin-Hadamard theorem concerning the distribution of primes;

(5) The Ikehara-Landau theorem and its application to the distribution of primes;

(6) The theorem connecting the mean of the square of the modulus of a function with a singular quadratic form in its integrated Fourier transform;

PREFACE xvii

(7) The theorem that a function which has a spectrum has a *positive* spectrum;

(8) A group of theorems concerning the spectra of linear transforms of a given function;

(9) The Weierstrass and Parseval theorems for almost periodic functions.

Naturally, most of these topics have already been treated in various monographs by myself and others. These are given in a bibliography at the end, and my own monographs will not necessarily be cited elsewhere.

After some deliberation, I decided to omit a discussion of harmonic analysis in the complex plane, on the ground that it required too much introductory material not needed elsewhere in the book, and would throw it out of balance.

The papers leading up to this book have been read in proof and most helpfully criticized by Professor J. D. Tamarkin of Brown University and by several of my colleagues and students. To these I wish to express my thanks. I wish to thank my Cambridge colleagues and the Syndics of the Cambridge University Press for causing this book to be written and making its publication possible. Particular gratitude is due to Professor Hardy of Cambridge and to Mr Skewes of the University of the Cape of Good Hope for their painstaking reading of the manuscript.

N. W.

Cambridge
July 1932

INTRODUCTION

§ 1. The Nature of Harmonic Analysis.

In the hierarchy of branches of mathematics, certain points are recognizable where there is a definite transition from one level of abstraction to a higher level. The first level of mathematical abstraction leads us to the concept of the individual numbers, as indicated for example by the Arabic numerals, without as yet any undetermined symbol representing some unspecified number. This is the stage of elementary arithmetic; in algebra we use undetermined literal symbols, but consider only individual specified combinations of these symbols. The next stage is that of analysis, and its fundamental notion is that of the arbitrary dependence of one number on another or of several others—the function. Still more sophisticated is that branch of mathematics in which the elementary concept is that of the transformation of one function into another, or, as it is also known, the operator. It is only in connection with the operational calculus that the true significance of harmonic analysis is to be appreciated.

Let us consider, then, an operation T, transforming functions $f(x)$ defined over $(-\infty, \infty)$ into functions $T\{f\} = g(x)$, defined over the same range. If

$$T\{f_1(x) + f_2(x)\} = T\{f_1(x)\} + T\{f_2(x)\};$$
$$T\{af(x)\} = aT\{f(x)\} \quad (a \text{ constant});$$

the operator $T\{f\}$ is said to be *linear*. Linear operations are common in physics—commoner indeed in physics than in nature, for the first approximation to a most definitely non-linear operator is often a linear one. An even more common type of operator has been given by Volterra the somewhat inappropriate name of an *operator of the closed cycle*. It has the property that if $T\{f\} = g$, and if $f(x+h) = f_1(x), g(x+h) = g_1(x)$, then

$$T\{f_1(x)\} = g_1(x).$$

If the argument x is taken to be the time, then most of the operators of physics are of the closed cycle, for there are extremely few physical processes in which a change in the time of commencement has any further effect than a corresponding change in the time at which any given stage of the process takes place.

The function e^{iux} plays a singularly important rôle with respect to operators of the closed cycle. This results from the fact that
$$e^{iu(x+h)} = e^{iuh} e^{iux},$$
or in words, that a time displacement of such a function produces no other change in it than to multiply it by a complex number of modulus 1. If the operator T which we are considering is both linear and of the closed cycle, it becomes a matter of interest to consider the result of applying it to linear combinations of functions such as e^{iux}. Let us note that
$$T(e^{iu(x+h)}) = T(e^{iuh} e^{iux}) = e^{iuh} T(e^{iux}),$$
so that if we put $T(e^{iux}) = \phi(x)$, we have
$$\phi(x+h) = e^{iuh} \phi(x),$$
or
$$\phi(h) = \phi(0) e^{iuh}.$$
Thus the effect of T on e^{iux} is merely to multiply it by a constant. It follows that

(1·1) $$T\left(\sum_1^N a_n e^{iu_n x}\right) = \sum_1^N a_n b_n e^{iu_n x},$$

where the b_n's depend only on u_n and T, not on a_n. In other words, if we regard the set of coefficients a_n as in some way representing the function $\sum_1^N a_n e^{iu_n x} = f(x)$, the linear operation T of the closed cycle applied to $f(x)$ corresponds to the multiplicative factors b_n applied to a_n. This fact and others that are its formal generalizations constitute the chief significance of methods of harmonic analysis.

In Chapter I our chief purpose is to extend theorems of the type (1·1) to the case where $\sum_1^N$ is replaced by an integral of the

INTRODUCTION

appropriate kind. We investigate some of the conditions under which a function $f(x)$ determines a function $g(u)$ given formally by

(1·2) $$g(u) = \frac{1}{\sqrt{2\pi}} \int_{-\infty}^{\infty} f(x) e^{-iux} dx,$$

and find that these conditions yield a result which we may formally write

(1·3) $$f(x) = \frac{1}{\sqrt{2\pi}} \int_{-\infty}^{\infty} g(u) e^{iux} dx,$$

which may be regarded as an extended form of $f(x) = \sum_{1}^{N} a_n e^{iu_n x}$. If we put

(1·4) $$g_1(u) = \frac{1}{\sqrt{2\pi}} \int_{-\infty}^{\infty} f_1(x) e^{-iux} dx,$$

we shall obtain

(1·5) $$\int_{-\infty}^{\infty} f_1(x - \xi) f(\xi) d\xi = \int_{-\infty}^{\infty} g_1(u) g(u) e^{iux} du.$$

The operator which turns $f(x)$ into $\int_{-\infty}^{\infty} f_1(x - \xi) f(\xi) d\xi$ thus corresponds formally to the operator of multiplication by $\sqrt{2\pi}\, g_1(u)$ applied to g.

In Chapter II we devote our attention to the asymptotic behaviour of such linear closed-cycle transforms of $f(x)$ as

(1·6) $$\int_{-\infty}^{\infty} K(x - \xi) f(\xi) d\xi.$$

This forms the subject-matter of the Tauberian theory—often in a form disguised by a change of variable, in which (1·6) is replaced by

$$\frac{1}{x} \int_{0}^{\infty} Q(x/\xi) \phi(\xi) d\xi.$$

It is not surprising, in view of the likeness of (1·6) to the left-hand expression in (1·5), that we shall find the function

$$\frac{1}{\sqrt{2\pi}} \int_{-\infty}^{\infty} K(x) e^{-iux} dx$$

playing an important rôle in this theory. This is all the more natural, in consideration of the fact that the asymptotic properties of a function are independent of any choice of origin, and are hence invariant under any transformation $x_1 = x + h$, which also leaves invariant all operators of the closed cycle.

In Chapter III we return to the consideration of expressions of a similar type to (1·3), but now adapted to the treatment of functions $f(x)$ which need not be small in some sense at infinity, as are those of Chapter I. The theory developed in Chapter II appears as a useful tool. One application of great importance is to the Bohr theory of almost periodic functions. We shall term the transformation which yields $f(x+\lambda)$ when applied to $f(x)$ (λ real) a *translation*. The conception of almost periodicity, like that of periodicity, involves no reference to an origin, and therefore has that type of invariance under translation which makes relevant the consideration of operators of the closed cycle and of an harmonic analysis in terms of the function e^{iux}.

§2. The Properties of the Lebesgue Integral.

It is well known that an adequate theory of the Fourier series can only be established on the basis of Lebesgue integration. All those theorems which proceed from a given function to its Fourier coefficients can indeed be established on the basis of any less inclusive concept, such as that of the Riemann integral, but the fundamental theorem which proceeds from a set of coefficients to the existence of a function having these Fourier coefficients, that of Riesz and Fischer, is simply false for any definition narrower than that of Lebesgue. In the theory of Fourier series, the function to be expanded and its coefficients are two very different sorts of things—the function is defined for a continuous infinity of values, but is periodic, and hence need only be given within a single period, whereas the coefficients are a discrete non-periodic set of numbers. In the case of the Fourier integral, however, if the function to be expanded is $f(x)$ of (1·3), the function $g(u)$ plays the part of the set of Fourier coefficients, and the formal similarity between (1·2) and (1·3) shows how

INTRODUCTION

impossible it is to say which of f and g is the function we are expanding and which is the set of coefficients. Both are non-periodic functions defined over continuous infinite ranges. Accordingly, it is impossible to segregate the difficulties of the theory of the Fourier integral into two categories, but instead we find that all the difficulties which in the Fourier series arise either on proceeding from the function to the coefficients or from the coefficients to the function are here met with in both arguments. It is totally impossible to establish a reasonably complete and symmetrical theory of the Fourier integral except on the basis of Lebesgue integration.

It is no part of the present book to attempt a thorough and systematic treatment of the Lebesgue integral. However, in view of the none too well-established position of the theory in the curriculum of English and American universities, it has seemed advisable to give a résumé of the definitions of Lebesgue measure and of the Lebesgue integral, together with a few of the cardinal theorems which come into constant use. In the choice of these theorems, I have been greatly influenced by the choice made for similar purposes by Professor Hardy in his lectures on Fourier series. It is my intention to make the selection full enough for this book to be intelligible to anyone who is fairly grounded in the elementary non-Lebesgue theory of the functions of a real variable. Naturally, none but the extremely indolent will content himself with taking the proof of these fundamental theorems on faith.

Let us start then with a set of points S on a finite interval (a, b). Let us inclose all points of S in a finite or denumerable set of intervals of total length M. The length M may of course be given different values by different choices of the set of inclosing intervals. We shall call the greatest lower bound of possible values of M the *outer measure* of S, and shall write it $\overline{m}(S)$.

Now let $\overline{S}$ be the set of all those points of (a, b) that do not belong to S. Let us put

$$\underline{m}(S) = b - a - \overline{m}(\overline{S}).$$

We shall term $\underline{m}(S)$ the *inner measure* of S. In case

$$\underline{m}(S) = \overline{m}(S),$$

we shall write $m(S)$ for the common value of these two quantities, shall term it the *measure* of S, and shall say that S is *measurable*. It is easy to show that the measure and the measurability of S do not depend on the length of the interval (a, b).

On an infinite interval, we shall term a set S measurable if whenever $a < b$, the portion of S in (a, b) is measurable. If we write $S(a, b)$ for this portion of S, we shall put

$$m(S) = \lim_{\substack{a \to -\infty \\ b \to \infty}} m(S(a, b)).$$

In two or more dimensions, the definition of measure is only changed in so far as we replace "interval" by "rectangle" or "rectangular parallelepiped" and "length" by "area" or "volume," in the sense appropriate to the number of dimensions in question. In the definition of the measure of a set in the whole plane or in all space, the interval (a, b) is replaced by an analogue, which is allowed to grow in all directions. The definitions of measure thus obtained are not restricted if we specify the orientation of our rectangles and rectangular parallelepipeds, but are independent of this orientation.

We shall define a *null set* as one of measure 0. A proposition involving a variable point is said to be true *almost everywhere* if it only fails to be true over (at most) a null set.

A function $f(x)$ is said to be *measurable* over (a, b) (a and b finite) if the set of values of x on (a, b) for which $\alpha \leqslant f(x) \leqslant \beta$ is measurable for every α and β. If the function $f(x)$ is k over the measurable set of points S lying within (a, b) and is zero elsewhere, we define the *Lebesgue integral*—we shall simply say *integral*—of $f(x)$ over (a, b) to be

(2·01) $$\int_a^b f(x)\, dx = k m(S).$$

If $f(x) = \sum_{1}^{n} f_k(x)$, where each one of the functions $f_k(x)$ is of

INTRODUCTION

the type for which the integral is defined in (2·01), we shall write
$$\int_a^b f(x)\,dx = \sum_1^n \int_a^b f_k(x)\,dx.$$
For any real bounded function of $f(x)$, we shall define
$$\overline{\int_a^b} f(x)\,dx,$$
the *upper integral* of $f(x)$, as the greatest lower bound of
$$\int_a^b g(x)\,dx,$$
where $g(x)$ is a function of the type for which the integral is defined in (2·01) and $g(x) \geqslant f(x)$ over (a, b). In the same circumstances, the lower integral is defined by
$$\underline{\int_a^b} f(x)\,dx = -\overline{\int_a^b}(-f(x))\,dx.$$
In case
$$\underline{\int_a^b} f(x)\,dx = \overline{\int_a^b} f(x)\,dx,$$
we shall write $\int_a^b f(x)\,dx$ for their common value.

We now introduce a pair of notations to which we shall have frequent occasion to return in later paragraphs. We put
$$f_{A,B}(x) = A\,(f(x) < A);\ = f(x)\,(A \leqslant f(x) < B);\ = B\,(B < f(x));$$
and
$$f_A(x) = f(x)\ (|f(x)| < A);\ = Af(x)/|f(x)|\ (|f(x)| > A).$$
Let us note that the second definition yields $f_{-A,A}(x)$ if $f(x)$ is real, but is significant even in the case of complex-valued functions. Let us further note for future reference that in all cases

(2·015) $\quad |f_{A,B}(x) - f_{A,B}(y)| \leqslant |f(x) - f(y)|;$
$\quad\quad\quad |f_A(x) - f_A(y)| \leqslant |f(x) - f(y)|.$

We now may complete the definition of Lebesgue integration for real unbounded functions. We define $\int_a^b f(x)\,dx$ by
$$\int_a^b f(x)\,dx = \lim_{\substack{A \to -\infty \\ B \to \infty}} \int_a^b f_{A,B}(x)\,dx.$$

In case the integral of $f(x)$ as thus defined exists, we shall say that $f(x)$ is integrable, or that it belongs to the class L_1.

Let us now enunciate certain propositions concerning measure and integration to which we shall have subsequent occasion to refer. The proofs are to be found in any standard treatise on the Lebesgue integral. We shall indicate these propositions by the letter X, as well as certain theorems which we shall prove but which belong rather to the background of the theories of this book than to the theories themselves.

X_1. *If S is the logical sum of the finite or denumerable set of non-overlapping measurable sets of points $S_1, S_2, \ldots, S_n, \ldots$, then S is measurable, and*
$$m(S) = m(S_1) + m(S_2) + \ldots + m(S_n) + \ldots.$$

X_2. *The logical sum and the logical product of a finite or denumerable set of measurable sets are measurable.*

X_3. *If $S_1, \ldots, S_n, \ldots$ is a sequence of measurable sets with logical product S, and S_k contains S_{k+1} for all values of k, then*
$$m(S) = \lim_{k \to \infty} m(S_k).$$

X_4. *If S_1 and S_2 are measurable sets, and S_1 contains S_2, then*
$$m(S_1) \geqslant m(S_2).$$

X_5. *The logical sum of a finite or denumerable set of null sets is a null set.*

X_6. *If $f(x)$ is a measurable function, so is $|f(x)|$, and if $f(x)$ and $g(x)$ are measurable functions, so are $f(x)g(x)$, and $f(x)/g(x)$, and $\alpha f(x) + \beta g(x)$, where α and β are any real constants, provided these functions are well-defined except on a null set.*

X_7. *The various definitions of Lebesgue integration given for different classes of functions are consistent, in the sense that where two are applicable, they yield the same value.*

X_8. *If $f(x)$ is a measurable function and $g(x)$ is an integrable function, and $|f(x)| \leqslant |g(x)|$ for all x, then $f(x)$ is integrable. In particular, all bounded measurable functions are integrable.*

INTRODUCTION

X$_9$. *If α and β are real constants, and $f(x)$ and $g(x)$ are integrable, so is $\alpha f(x) + \beta g(x)$. We have*

$$\int_a^b [\alpha f(x) + \beta g(x)]\, dx = \alpha \int_a^b f(x)\, dx + \beta \int_a^b g(x)\, dx.$$

X$_{10}$. *If $f(x)$ is a non-negative integrable function,*

$$\int_a^b f(x)\, dx \geq 0.$$

As corollaries, we have for any integrable f,

$$\int_a^b |f(x)|\, \geq 0;$$

for $f \geq g$,
$$\int_a^b f(x)\, dx \geq \int_a^b g(x)\, dx;$$

and in general

$$\left| \int_a^b f(x)\, dx \right| \leq \int_a^b |f(x)|\, dx \leq |b-a| \limsup_{a \leq x \leq b} |f(x)|.$$

X$_{11}$. *If $\{f_n(x)\}$ is a sequence of integrable functions, if*

$$f_n(x) \leq f_{n+1}(x)$$

for every x on (a, b) with the possible exception of a null set, and if

$$I = \lim_{n \to \infty} \int_a^b f_n(x)\, dx$$

is finite, then, with the possible exception of a null set of values of x, the limit

$$f(x) = \lim_{n \to \infty} f_n(x)$$

exists and is finite, and

$$I = \int_a^b f(x)\, dx.$$

As a corollary, if $\overset{\infty}{\underset{1}{\Sigma}} g_n(x)$ is a series of positive integrable functions, it may be integrated term by term, in the sense that if

$$\sum_1^\infty \int_a^b g_n(x)\, dx$$

converges, then $\sum\limits_{1}^{\infty} g_n(x)$ converges almost everywhere, and

(2·02) $$\int_a^b \left[\sum_{1}^{\infty} g_n(x) \right] dx = \sum_{1}^{\infty} \int_a^b g_n(x)\, dx.$$

This is known as the test of *monotone convergence* for the termwise integrability of a series.

X_{12}. *If $\{f_n(x)\}$ is a sequence of integrable functions, if there exists an integrable function $F(x)$ independent of n such that*

$$|f_n(x)| \leqslant F(x)$$

for all values of x on (a, b), and if

$$f(x) = \lim_{n \to \infty} f_n(x)$$

almost everywhere, then $f(x)$ is integrable, and

$$\int_a^b f(x)\, dx = \lim_{n \to \infty} \int_a^b f_n(x)\, dx.$$

In particular, $F(x)$ may be a constant. Applying this result to series, we see that if $\sum\limits_{1}^{\infty} g_n(x)$ is a series of integrable functions with uniformly bounded partial sums, or with partial sums uniformly dominated by an integrable function $F(x)$, and if it converges almost everywhere, then it may be integrated term by term, in the sense that both sides of (2·02) will exist and will have the same value. This is known as the test of *dominated convergence* for the term-by-term integrability of a series, or in the case where $F(x)$ is constant, the test of *bounded convergence*.

X_{13}. *If $f(x)$ is integrable, $\int_a^x f(\xi)\, d\xi$ is a continuous function of its upper limit of integration x. We have almost everywhere*

$$f(x) = \frac{d}{dx} \int_a^x f(\xi)\, d\xi.$$

This is known as the fundamental theorem of the calculus. Another form of it asserts that

$$\frac{1}{\epsilon} \int_x^{x+\epsilon} f(\xi)\, d\xi \to f(x)$$

almost everywhere as $\epsilon \to 0$.

INTRODUCTION

X_{14}. *If S is a measurable set of points on the finite interval (a, b) and ϵ is any positive quantity, then there is a set of points S_1, consisting of a finite number of finite intervals, and such that the measure of the set of points in one but not both of the sets of points S and S_1 is less than ϵ.*

X_{15}. *If $f(x)$ is integrable, and ϵ is any positive quantity, there is a measurable function $f_1(x)$, assuming only a finite number of values, such that*

(2·03) $$\int_a^b |f(x) - f_1(x)|\, dx < \epsilon.$$

This results from the fact that we may take A so large that

(2·04) $$\int_a^b |f(x) - f_A(x)|\, dx < \tfrac{1}{2}\epsilon,$$

and that we may divide the interval $(-A, A)$ into sub-intervals (A_n, A_{n+1}), so small that for each A_n

$$(b-a)(A_{n+1} - A_n) < \tfrac{1}{2}\epsilon.$$

If we now put
$$f_1(x) = -A \quad (f(x) < -A); \quad f_1(x) = A_n \quad (A_n \leqslant f(x) < A_{n+1});$$
$$f_1(x) = A \quad (A \leqslant f(x));$$
it will readily follow from X_{10} that

$$\int_a^b |f_1(x) - f_A(x)| < \tfrac{1}{2}\epsilon,$$

and this and (2·04) give (2·03).

X_{16}. *If $f(x)$ is integrable, and ϵ is any positive quantity, there is a function $f_2(x)$, equal to a constant over each of a finite number of intervals which constitute collectively (a, b), and such that*

(2·05) $$\int_a^b |f(x) - f_2(x)|\, dx < \epsilon.$$

In view of X_{15}, we may assume that $f(x)$ takes only a finite number of distinct values. Such a function is of the form

$$\sum_1^N a_k g_k(x),$$

where each $g_k(x)$ is unity over a measurable set and is elsewhere zero. Proposition X_{10} shows that the integral of the modulus of a sum does not exceed the sum of the integrals of the modulus of the summands. Thus if we establish X_{16} for the particular case where $f(x)$ assumes the value 1 over a measurable set and is elsewhere 0, we shall have established it for the g_k's, and consequently for all functions $\sum_1^N a_k g_k(x)$, which we have seen to be equivalent to establishing it for all integrable functions. In the case where $f(x)$ assumes the value 1 over a measurable set and is elsewhere 0, however, X_{16} reduces to X_{14}, which we have asserted.

X_{17}. *If $f(x)$ is integrable,*

$$(2·06) \qquad \lim_{\eta \to 0} \int_a^{b-\eta} |f(x+\eta) - f(x)| \, dx = 0.$$

To begin with, let $f_2(x)$ be a function equal to a constant over each of a finite number of intervals collectively filling (a, b), or as we shall say from now on, let it be a *step-function*. Let (2·05) be true. By proposition X_{16}, such a step-function $f_2(x)$ will exist. The function $f_2(x)$ will be of limited total variation, say V. Then, if η is small enough,

$$\int_a^{b-\eta} |f_2(x+\eta) - f_2(x)| \, dx = V\eta.$$

Hence

$$(2·07) \qquad \lim_{\eta \to 0} \int_a^{b-\eta} |f_2(x+\eta) - f_2(x)| \, dx = 0.$$

It follows from (2·05) that

$$\int_a^{b-\eta} |f(x+\eta) - f(x)| \, dx - \int_a^{b-\eta} |f_2(x+\eta) - f_2(x)| \, dx$$

$$\leq \int_a^{b-\eta} |f(x+\eta) - f(x) - f_2(x+\eta) + f_2(x)| \, dx$$

$$\leq \int_a^{b-\eta} |f(x+\eta) - f_2(x+\eta)| \, dx + \int_a^{b-\eta} |f(x) - f_2(x)| \, dx$$

$$= \int_{a+\eta}^{b} |f(x) - f_2(x)| \, dx + \int_a^{b-\eta} |f(x) - f_2(x)| \, dx$$

$$\leq 2 \int_a^b |f(x) - f_2(x)| \, dx = 2\epsilon.$$

Thus, by (2·07),

$$\overline{\lim_{\eta \to 0}} \int_a^{b-\eta} |f(x+\eta) - f(x)| \, dx \leq 2\epsilon,$$

and since ϵ is arbitrarily small, (2·06) will follow.

Let us now turn to the definition of the Lebesgue integral over an infinite interval. We shall say that $f(x)$ is integrable (or belongs to L_1) over $(-\infty, \infty)$ if

(2·08) $$\lim_{\substack{a \to -\infty \\ b \to \infty}} \int_a^b |f(x)| \, dx$$

exists as a finite quantity. We shall then write

(2·09) $$\int_{-\infty}^{\infty} f(x) \, dx = \lim_{\substack{a \to -\infty \\ b \to \infty}} \int_a^b f(x) \, dx.$$

Let it be noted that a Lebesgue integral is always an absolutely convergent integral. Whether (2·08) exists or not, (2·09) is said to define $\int_{-\infty}^{\infty} f(x) \, dx$ as a *Cauchy integral*.

It is a trivial remark that, if $f(x)$ vanishes for all x outside (a, b), then

$$\int_{-\infty}^{\infty} f(x) \, dx = \int_a^b f(x) \, dx.$$

Propositions $X_8, X_9, X_{10}, X_{11}, X_{12}, X_{13}$ remain true when (a, b) is replaced by $(-\infty, \infty)$. It is worth noting that the integration of series term by term is still possible under the criteria of monotone or of dominated convergence, but not under that of bounded convergence, as a positive constant is not summable over $(-\infty, \infty)$. The definition and properties of integrals with one limit infinite scarcely need a separate discussion.

If $f(x)$ belongs to L_1,

$$\lim_{a \to -\infty} \int_{-\infty}^a f(x) \, dx = 0, \quad \lim_{a \to \infty} \int_b^\infty f(x) \, dx = 0.$$

With this in view, it is easy to see that X_{15} still holds over $(-\infty, \infty)$, and as a consequence, X_{16}. X_{17} now becomes:

X_{18}. *If $f(x)$ is integrable over $(-\infty, \infty)$,*
$$\lim_{\eta \to 0} \int_{-\infty}^{\infty} |f(x+\eta) - f(x)|\, dx = 0.$$

If $f(x)$ is a function with complex values but a real argument, we may write it in the form $f_1(x) + if_2(x)$ with real f_1 and f_2. We shall then define its integral by
$$\int_a^b f(x)\, dx = \int_a^b f_1(x)\, dx + i \int_a^b f_2(x)\, dx,$$
with a similar definition in the case where a or b is infinite, or they both are infinite.

An important theorem is that of Riemann-Lebesgue, to the effect that:

X_{19}. *If $f(x)$ is integrable over $(-\infty, \infty)$,*
$$\lim_{u \to \pm\infty} \int_{-\infty}^{\infty} f(x) e^{iux}\, dx = 0.$$

To establish this, let us notice that
$$\int_{-\infty}^{\infty} f(x) e^{iux}\, dx = \frac{1}{2} \int_{-\infty}^{\infty} f(x) \left[e^{iux} - e^{iu\left(x - \frac{\pi}{u}\right)} \right] dx$$
$$= \frac{1}{2} \int_{-\infty}^{\infty} \left[f(x) - f\left(x - \frac{\pi}{u}\right) \right] e^{iux}\, dx$$
$$\leq \frac{1}{2} \int_{-\infty}^{\infty} \left| f\left(x + \frac{\pi}{u}\right) - f(x) \right| dx,$$
so that the result follows from X_{18}.

In the proof of X_{19}, we have tacitly used a special case of the following proposition.

X_{20}. *We have*
$$\int_{f(a)}^{f(b)} F(f(x)) f'(x)\, dx = \int_a^b F(x)\, dx,$$
whenever $f(x)$ is the integral of its derivative, which is non-negative, and F is integrable. This theorem is true over a finite or infinite interval. It includes the fact that $F(f(x)) f'(x)$ is summable. This is also related to:

X_{21}. *A measurable function of a measurable function is measurable.* Together with X_6, this guarantees that all functions

INTRODUCTION 15

which arise in a normal manner in operations with measurable functions are measurable. Accordingly, we shall assume all functions occurring in this book to be measurable, unless the opposite is explicitly stated, and shall only prove the measurability of a function arising in the course of our argument, in case the demonstration offers some real difficulty.

Another proposition which it is as well to state explicitly is:

X_{22}. *When applied to a continuous function, the Lebesgue integral is identical with the classical Riemann integral. This is also true when the function is made up of a finite number of continuous pieces.*

This allows us to employ the classical integrals of the tables of integration. Of course, the Riemann integral is equivalent to the Lebesgue integral over a much wider range, embracing all absolutely convergent integrals. However, the Riemann integral is of relatively little importance in the theory of Fourier series and integrals, save as the classical definition applying to continuous and "step-wise continuous" functions.

A further justification of familiar procedure is:

X_{23}. *If $f(x)$ and $g(x)$ are integrals of their derivatives,*

$$\int_a^b f(x) g'(x) dx = f(b) g(b) - f(a) g(a) - \int_a^b g(x) f'(x) dx.$$

We now come to the integral of a function of two or more variables. If S is a set of points, let us say in two dimensions, the definition and properties of $\iint_S f(x, y) \, dx \, dy$ do not differ, *mutatis mutandis*, from that of the simple integral. Summability receives a precisely analogous definition. With the exception of X_{13} and subsequent propositions, where the enunciation needs a little revision, all the propositions so far stated are independent of dimensionality.

Indeed, the relation between integration in one dimension and integration in a space of more than one dimension is closer than an analogy, and amounts to an isomorphism, or practically to an identity. Let us consider the following mapping of the square $(0 \leqslant x \leqslant 1,\ 0 \leqslant y \leqslant 1)$ on the linear segment $(0 \leqslant y \leqslant 1)$: if x has the representation

(2·10) $\qquad \cdot x_1 x_2 x_3 \ldots x_n \ldots$

in the binary scale of notation, and y the representation

(2·11) $\qquad \cdot y_1 y_2 y_3 \ldots y_n \ldots,$

then z is to have the representation

(2·12) $\qquad \cdot x_1 y_1 x_2 y_2 \ldots x_n y_n \ldots$

obtained by alternately choosing digits from x and y in their proper order. This mapping will be one-one unless x or y or z admits a terminating binary representation other than 0 or 1. If one and only one of the two numbers x and y admits such a terminating representation, the corresponding point on the square is the image of two distinct points of the z-line, each of which corresponds to a single point of the plane. If both the numbers x and y admit terminating representations other than 1 or 0, (x, y) may be the image of as many as four distinct points of the z-line. In these circumstances, the same point on the z-line may correspond to two distinct points in the plane. These exceptional values of x and y form a set of zero measure in x and in y, and the corresponding points of the square are of zero plane measure. The points of z corresponding to the exceptional values of x or to the exceptional values of y are likewise a set of zero linear measure. Thus, with the exception of a null set on the square and a corresponding null set on the z-line, the mapping given by (2·12) is one-one.

Let S be any set of points on the square, and let Σ be the set of its image-points on the z-line. It may be shown that the plane measure of S is equal to the linear measure of Σ. It is clearly a matter of indifference whether we count points with multiple images once, a number of times equal to their multiplicity, or not at all, since they form a null set as well as do their images, and do not contribute either to the measure of S or to that of Σ.

We may write the mapping indicated by (2·10—12) in the form

(2·13) $\qquad x = x(z),\quad y = y(z).$

The functions $x(z)$ and $y(z)$ will, however, not be continuous, as we can see by considering terminating values of z such as ·1 = 0·111.... Here $x(z)$ has the single value ·1 = ·01111..., but $y(z)$ assumes the two values 0 and ·1111... = 1. If z is a non-terminating binary number slightly smaller than ·1, $y(z)$ will be close to 1, while if z is a non-terminating binary slightly greater than ·1, $y(z)$ will be close to 0. On the other hand, $x(z)$ and $y(z)$ will be continuous at every value of z which has not a terminating binary representation.

There are, however, pairs of continuous functions $x(z)$ and $y(z)$ which generate the same mapping of measure as do the functions of (2·13). Examples of such functions have been given by Peano*, in connection with his space-filling curves. In exactly the same way, it is possible to find three continuous functions $x(u)$, $y(u)$, and $z(u)$, which map the segment $0 \leqslant u \leqslant 1$ on the cube $0 \leqslant x \leqslant 1$, $0 \leqslant y \leqslant 1$, $0 \leqslant z \leqslant 1$, with preservation of measure, and so for any finite or even infinite† number of dimensions. Under such a mapping, the two-dimensional or higher integral is the precise translation of the integral in a single dimension. In the case of two dimensions, for example,

$$\int_0^1 \int_0^1 f(x, y) \, dx \, dy = \int_0^1 f(x(z), y(z)) \, dz,$$

in the sense that when either exists as a Lebesgue integral, the other exists and assumes the same value. Thus *no property of the Lebesgue integral not explicitly involving a specified number of dimensions in its enunciation is restricted to integrals in any particular number of dimensions.* This is also true over an infinite range of integration.

There are, however, propositions concerning integration in space of more than one dimension which are essentially new, and which concern the relations between integrals of different dimensionality. Perhaps the most convenient way to state them is to regard all integrals as integrals over an infinite range, and to

* Curves of this sort are discussed and represented in Hobson, *Theory of the Functions of a Real Variable*, vol. I, §§ 326–328.

† Cf. Wiener 1. (Citations in this form refer to the Bibliography.)

avoid any restriction of the region of integration by the insertion of an integrand 0 everywhere outside the specified finite range. It will be remembered that all our integrals are by definition absolutely convergent.

A particular theorem of this type is:

X$_{24}$. *If $f(x, y)$ is summable, in (x, y), then it is summable in x for almost all y, and if P represents the entire (x, y) plane,*

$$\iint_P f(x, y)\, dx\, dy = \int_{-\infty}^{\infty} dy \int_{-\infty}^{\infty} f(x, y)\, dx.$$

X$_{25}$. *If $f(x, y)$ is measurable, if $|f(x, y)| \leq F(x, y)$ for all arguments, and if any one of these three integrals*

$$\iint_P F(x, y)\, dx\, dy,$$

$$\int_{-\infty}^{\infty} dy \int_{-\infty}^{\infty} F(x, y)\, dx,$$

$$\int_{-\infty}^{\infty} dx \int_{-\infty}^{\infty} F(x, y)\, dy$$

exists, then all the following integrals exist, and are identical:

$$\iint_P f(x, y)\, dx\, dy,$$

$$\int_{-\infty}^{\infty} dy \int_{-\infty}^{\infty} f(x, y)\, dx,$$

$$\int_{-\infty}^{\infty} dx \int_{-\infty}^{\infty} f(x, y)\, dy.$$

We now introduce the class of functions L_2, consisting of all measurable functions $f(x)$ for which

$$\int_a^b |f(x)|^2\, dx < \infty.$$

Let it be noted that if (a, b) is a finite range, L_2 may readily be shown to include L_1, as if $|f(x)|$ is large, $|f(x)|^2$ is even larger. On an infinite range this is no longer true, since an integral over an infinite range can be large either in virtue of the largeness of the integrand over a finite region, or of the slow decrease of the integrand over an infinite region. There is no relation of inclusion between L_1 and L_2 over $(-\infty, \infty)$. It is also not true that the proposition, "f^2 is summable," implies, "f belongs to L_2," since it does not even imply that f is measurable.

One of the most important theorems of analysis is that of the Schwarz inequality. This asserts:

X$_{26}$. *If $f(x)$ and $g(x)$ belong to L_2, $f(x)g(x)$ belongs to L_1, and*

(2·14) $$\left[\int_a^b |f(x)g(x)|\,dx\right]^2 \leqslant \int_a^b |f(x)|^2\,dx \int_a^b |g(x)|^2\,dx.$$

To establish this, let us first notice that $f(x)g(x)$ is measurable, and that
$$2|f(x)g(x)| \leqslant |f(x)|^2 + |g(x)|^2.$$
Thus $f(x)g(x)$ is less in modulus than a summable function, and is summable. From this it follows that
$$\{|f(x)| + \lambda |g(x)|\}^2$$
is summable, which gives us
$$0 \leqslant \int_a^b \{|f(x)| + \lambda |g(x)|\}^2 dx$$
$$= \int_a^b |f(x)|^2\,dx + 2\lambda \int_a^b |f(x)g(x)|\,dx + \lambda^2 \int_a^b |g(x)|^2\,dx.$$
This is thus a definite quadratic polynomial in λ, and its discriminant cannot be positive. This discriminant is, however,
$$4\left\{\left[\int_a^b |f(x)g(x)|\,dx\right]^2 - \int_a^b |f(x)|^2\,dx \int_a^b |g(x)|^2\,dx\right\},$$
and (2·14) is established. Of course, (a, b) need not be finite.

Let it be noted that an exactly similar argument will show that if $\sum_1^\infty |a_n^2|$ and $\sum_1^\infty |b_n^2|$ converge, $\sum_1^\infty |a_n b_n|$ converges, and
$$\left\{\sum_1^\infty |a_n b_n|\right\}^2 \leqslant \sum_1^\infty |a_n^2| \sum_1^\infty |b_n^2|.$$

The Schwarz inequality is a particular case of a more general inequality—that of Hölder*. Though we shall not use the

---

* The proof here given is that which Hardy (*Journal London Math. Soc.* 4 (67-68), 5 (80)) attributes to Littlewood and F. Riesz (*Bollettino dell' Unione Mat. Italiana*, 7 (1928), 77-79).

Hölder inequality in the present book, we shall state it and prove it for the sake of completeness.

X_{26}'. *If $p>1$ and $1/p + 1/q = 1$, and if $\int_a^b |f(x)|^p\, dx$ and $\int_a^b |g(x)|^q\, dx$ are finite, then $\int_a^b f(x)\, g(x)\, dx$ exists, and*

(2·15)
$$\left| \int_a^b f(x)\, g(x)\, dx \right| \leqslant \left[\int_a^b |f(x)|^p\, dx \right]^{1/p} \left[\int_a^b |g(x)|^q\, dx \right]^{1/q}.$$

There is no real restriction in (2·15) if we take f and g positive, and suppose

(2·16)
$$\int_a^b [f(x)]^p\, dx = 1, \quad \int_a^b [g(x)]^q\, dx = 1.$$

We then must show that
$$\int_a^b f(x)\, g(x)\, dx \leqslant 1.$$

For this we need a lemma to the effect that if α and β are non-negative and $0 \leqslant \lambda \leqslant 1$, then

(2·17) $\qquad \alpha^\lambda \beta^{1-\lambda} \leqslant \lambda \alpha + (1-\lambda)\beta.$

We may assume without restriction that $\alpha \geqslant \beta$, and, as $\alpha = \beta$ is trivial, that $\alpha > \beta$. Since (2·17) is homogeneous, we may take $\beta = 1$. Proposition (2·17) then becomes

(2·18) $\qquad \alpha^\lambda \leqslant 1 + \lambda(\alpha - 1).$

However, by Taylor's theorem with remainder,
$$\alpha^\lambda = [1+(\alpha-1)]^\lambda = 1 + \lambda(\alpha-1) + \frac{\lambda(\lambda-1)}{2}\theta(\alpha-1)^2,$$
where θ lies between 0 and 1. As $0 \leqslant \lambda \leqslant 1$, (2·18) results immediately.

It follows from (2·17) that
$$f(x)\, g(x) \leqslant \frac{1}{p}[f(x)]^p + \frac{1}{q}[g(x)]^q,$$
and hence by (2·16) that
$$\int_a^b f(x)\, g(x)\, dx \leqslant \frac{1}{p} + \frac{1}{q} = 1.$$

This establishes X$_{26}$'. It will be noted that the range (a, b) may be finite or infinite.

If we put $p = q = 2$, the Hölder inequality reduces to the Schwarz inequality.

Another general inequality is that of Minkowski*. This asserts:

X$_{27}$. *Let $p > 1$, and let $|f(x)|^p$ and $|g(x)|^p$ belong to L. Then $|f(x) + g(x)|^p$ belongs to L, and*

$$(2\cdot 19) \quad \left\{\int_a^b |f(x) + g(x)|^p\right\}^{1/p} \leq \left\{\int_a^b |f(x)|^p\right\}^{1/p} + \left\{\int_a^b |g(x)|^p\right\}^{1/p}.$$

Here the sign of equality is only to be taken if there is a μ such that $f(x) = \mu g(x)$ almost everywhere.

Here we need the inequality that if $0 \leq \lambda \leq 1$, $\alpha \geq 0$, $p > 1$, then

$$(2\cdot 20) \quad \{1 + \lambda(\alpha - 1)\}^p \leq 1 + \lambda(\alpha^p - 1).$$

To prove (2·20), let us notice that if $\lambda = 0$ or $\lambda = 1$, the two sides of (2·20) are identical. Let us consider the first and second derivatives of the expression

$$\Phi(\lambda) = 1 + \lambda(\alpha^p - 1) - \{1 + \lambda(\alpha - 1)\}^p.$$

These are
$$\Phi'(\lambda) = \alpha^p - 1 - p\{1 + \lambda(\alpha - 1)\}^{p-1}(\alpha - 1),$$
and
$$\Phi''(\lambda) = -p(p - 1)(\alpha - 1)^2\{1 + \lambda(\alpha - 1)\}^{p-2}.$$

We see that $\Phi''(\lambda)$ is negative over $(0, 1)$. Thus $\Phi(\lambda)$ is concave downward over $(0, 1)$, and vanishes at 0 and 1, from which it results that $\Phi(\lambda) \geq 0$ over the interval in question. This is however equivalent to (2·20).

Let us note that unless $\alpha = 1$, $\Phi''(\lambda)$ is properly positive, and the sign of equality in (2·20) is only possible for $\lambda = 0$ or $\lambda = 1$.

From (2·20) we readily deduce that if $0 \leq \lambda \leq 1$, $\phi \geq 0$, $\psi \geq 0$, $p > 1$, then
$$\{\lambda\phi + (1 - \lambda)\psi\}^p \leq \lambda\phi^p + (1 - \lambda)\psi^p,$$

---

* Hardy (*loc. cit.*) gives a reduction, due to F. Riesz, of the Minkowski inequality to the Hölder inequality.

with equality only when $\lambda = 0$ or 1 or $\phi = \psi$. Thus if over (a, b) $\phi(x) \geqslant 0$, $\psi(x) \geqslant 0$, and if $0 \leqslant \lambda \leqslant 1$, $p > 1$,
$$\int_a^b [\phi(x)]^p \, dx = \int_a^b [\psi(x)]^p \, dx = 1,$$
we obtain $\int_a^b \{\lambda \phi(x) + (1-\lambda) \psi(x)\}^p \, dx \leqslant 1.$
Thus
(2·21) $\left\{\int_a^b \{\lambda \phi(x) + (1-\lambda) \psi(x)\}^p \, dx\right\}^{1/p} \leqslant \lambda + (1-\lambda)$
$= \left\{\int_a^b [\lambda \phi(x)]^p \, dx\right\}^{1/p} + \left\{\int_a^b [(1-\lambda) \psi(x)]^p \, dx\right\}^{1/p}.$

Now let $\phi(x) = |f(x)| \Big/ \left\{\int_a^b |f(x)|^p \, dx\right\}^{1/p};$
$\psi(x) = |g(x)| \Big/ \left\{\int_a^b |g(x)|^p \, dx\right\}^{1/p};$
$\lambda = \left\{1 + \left[\int_a^b |g(x)|^p \, dx \Big/ \int_a^b |f(x)|^p \, dx\right]^{1/p}\right\}^{-1}.$

Then
$\lambda \phi(x) = |f(x)| \Big/ \left\{\left[\int_a^b |f(x)|^p \, dx\right]^{1/p} + \left[\int_a^b |g(x)|^p \, dx\right]^{1/p}\right\};$
$(1-\lambda) \psi(x) = |g(x)| \Big/ \left\{\left[\int_a^b |f(x)|^p \, dx\right]^{1/p} + \left[\int_a^b |g(x)|^p \, dx\right]^{1/p}\right\}.$

If we substitute these values in (2·21), and remember that (2·19) is homogeneous in f and g, we obtain (2·19). It will be seen that the sign of equality in (2·19) is only possible when $f(x) \equiv 0$ or $g(x) \equiv 0$ or $\phi(x) \equiv \psi(x)$, the identity being interpreted as identity almost everywhere. In each such case there is a μ such that $f(x) = \mu g(x)$ almost everywhere.

We shall only use in this book that particular case of the Minkowski inequality for which $p = 2$. This reads:

X_{27}'. *If $f(x)$ and $g(x)$ belong to L_2, so does $f(x) + g(x)$, and*
(2·22)
$$\left\{\int_a^b |f(x) + g(x)|^2 \, dx\right\}^{\frac{1}{2}} \leqslant \left\{\int_a^b |f(x)|^2 \, dx\right\}^{\frac{1}{2}} + \left\{\int_a^b |g(x)|^2 \, dx\right\}^{\frac{1}{2}}.$$

We may prove this directly with somewhat simpler arguments than we have used in the case of X_{27}. Let us notice that both

INTRODUCTION

sides of (2·22) may be squared without affecting its validity. If we remove terms common to both sides, (2·22) is equivalent to

$$(2\cdot 23) \quad 2R\int_a^b f(x)\bar{g}(x)\,dx \leq 2\left\{\int_a^b |f(x)|^2\,dx \int_a^b |g(x)|^2\,dx\right\}^{\frac{1}{2}},$$

where the bar over a symbol is used here, as always in this book, for the conjugate complex quantity. However, we know that

$$R\int_a^b f(x)\bar{g}(x)\,dx \leq \left|\int_a^b f(x)\bar{g}(x)\,dx\right| \leq \int_a^b |f(x)\,g(x)|\,dx,$$

so that (2·23) is an immediate consequence of (2·14). Here too the interval (a, b) may be infinite, and here too the analogous proposition is true, that if $\sum_1^\infty |a_n^2|$ and $\sum_1^\infty |b_n^2|$ converge, so does $\sum_1^\infty |(a_n + b_n)^2|$, and

$$\left\{\sum_1^\infty |(a_n + b_n)^2|\right\}^{\frac{1}{2}} \leq \left\{\sum_1^\infty |a_n^2|\right\}^{\frac{1}{2}} + \left\{\sum_1^\infty |b_n^2|\right\}^{\frac{1}{2}}.$$

When the infinite series are replaced by finite sums and the a's and b's are real, this proposition merely asserts that no side of a triangle can exceed in length the sum of the remaining two sides.

Proposition X_{16} asserted that it is possible to approximate to any function of L_1 by a step-function. The corresponding proposition is true of a function of L_2. To see this we need first:

X_{28}. *If $f(x)$ belongs to L_2 over $(-\infty, \infty)$*

$$\lim_{A \to \infty} \int_{-\infty}^\infty |f_A(x) - f(x)|^2\,dx = 0.$$

This may be proved by dominated convergence, inasmuch as $|f_A(x) - f(x)|^2$ tends to zero for every x outside a certain null set, and is nowhere greater than $|f(x)|^2$.

If we put

$$_A f(x) = f(x) \quad (|x| < A); \quad = 0 \quad (|x| \geq A),$$

a trivial theorem is

X_{29}. *If $f(x)$ belongs to L_2,*

$$\lim_{A \to \infty} \int_{-\infty}^\infty |_A f(x) - f(x)|^2\,dx = 0.$$

If we combine this with X_{28} and the Minkowski inequality, we get

$$\lim_{A \to \infty} \int_{-\infty}^{\infty} |{}_A f_A(x) - f(x)|^2 \, dx = 0.$$

We have already shown in X_{16} that there is a step-function $f_2(x)$ such that

$$\int_{-\infty}^{\infty} |{}_A f_A(x) - f_2(x)| \, dx < \epsilon,$$

and we can clearly take $|f_2(x)| \leqslant A$. Now,

$$\int_{-\infty}^{\infty} |{}_A f_A(x) - f_2(x)|^2 \, dx \leqslant 2A \int_{-\infty}^{\infty} |{}_A f_A(x) - f_2(x)| \, dx \leqslant 2A\epsilon,$$

so that if $\eta > 0$, we may first choose A so large that

$$\int_{-\infty}^{\infty} |{}_A f_A(x) - f(x)|^2 \, dx < \tfrac{1}{4}\eta,$$

and then determine a step-function $f_2(x)$, such that

$$\int_{-\infty}^{\infty} |{}_A f_A(x) - f_2(x)|^2 \, dx < \tfrac{1}{4}\eta.$$

Then, by the Minkowski inequality,

(2·24) $$\int_{-\infty}^{\infty} |f(x) - f_2(x)|^2 \, dx < \eta.$$

We have thus proved

X_{30}. *If $f(x)$ belongs to L_2, and $\eta > 0$, there exists a step-function $f_2(x)$ for which (2·24) is true.*

We may now prove the following proposition in a manner quite analogous to that in which we proved X_{17}:

X_{31}. *If $f(x)$ belongs to L_2,*

$$\lim_{\eta \to 0} \int_{-\infty}^{\infty} |f(x+\eta) - f(x)|^2 \, dx = 0.$$

Another idea which we need to introduce as a working tool is that of the Stieltjes integral. We shall not need the complete theory of the Lebesgue-Stieltjes integral, but shall be able to confine our attention to the more elementary Stieltjes

INTRODUCTION 25

integral analogous to that of Riemann. If $f(x)$ is a uniformly continuous function over (a, b) and $g(x)$ a function of limited total variation, we define $\int_a^b f(x)\, dg(x)$ as

$$\lim \Sigma f(\xi_k)\, [g(x_{k+1}) - g(x_k)],$$

where $\quad a = x_0 < x_1 < \ldots < x_n = b; \quad x_k \leqslant \xi_k \leqslant x_{k+1}$,

and the limit is taken as the maximum of the quantities

$$x_{k+1} - x_k$$

tends to zero by repeated subdivision of the interval (a, b). This definition yields an integral independent of the choice of the quantities x_k and ξ_k, and the particular method of procedure to the limit. Important theorems which we shall use without proof are:

X_{32}. *If $f(x)$ is a continuous function with maximum modulus F, and $g(x)$ is a function of limited total variation V over (a, b), then*

$$\left| \int_a^b f(x)\, dg(x) \right| \leqslant FV.$$

X_{33}. *If $g(x)$ is of limited total variation and the integral of its derivative, and $f(x)$ is continuous,*

$$\int_a^b f(x)\, dg(x) = \int_a^b f(x)\, \frac{dg(x)}{dx}\, dx.$$

X_{34}. *If $\phi(x)$ is any monotone continuous function of x, and f and g are as in the definition of a Stieltjes integral,*

$$\int_{\phi(a)}^{\phi(b)} f(x)\, dg(x) = \int_a^b f(\phi(x))\, dg(\phi(x)).$$

X_{35}. *If $f(x)$ is continuous, and $g(x)$ is a step-function, then*

$$\int_a^b f(x)\, dg(x) = \sum_1^N f(x_n)\, [g(x_n + 0) - g(x_n - 0)],$$

where the points x_n are the points of discontinuity of $g(x)$ over the closed interval (a, b), taking $g(a - 0) = g(a); \; g(b + 0) = g(b)$.

X₃₆. *If $f(x)$ is the integral of its derivative and $g(x)$ is of limited total variation,*

$$\int_a^b f(x)\,dg(x) = f(b)g(b) - f(a)g(a) - \int_a^b g(x)f'(x)\,dx.$$

X₃₇. *If $\phi(s)$ is continuous over (a, b), if $\gamma(x)$ is of limited variation over (c, d), and if $\alpha(x, s)$ is continuous in x for all values of s in (a, b), and of uniformly limited variation in s for all values of x in (c, d), then the integrals*

$$\int_c^d \left[\int_a^b \phi(s)\,d_s\alpha(x, s)\right] d\gamma(x), \quad \int_a^b \phi(s)\,d_s \int_c^d \alpha(x, s)\,d\gamma(x)$$

exist and are equal*.

If we make use of the principle that whenever $F(x, y)$ is an increasing function of x for every y, and an increasing function of y for every x,

$$\lim_{x \to \infty} \lim_{y \to \infty} F(x, y) = \lim_{y \to \infty} \lim_{x \to \infty} F(x, y) = \lim_{x, y \to \infty} F(x, y),$$

in the sense that whenever one of these limits exists, the others exist and have the same value, and if we put as a definition

$$\int_{-\infty}^{\infty} f(x)\,dg(x) = \lim_{\substack{a \to -\infty \\ b \to \infty}} \int_a^b f(x)\,dg(x),$$

then we may deduce as a corollary of X₃₇:

X₃₈. *Let the hypothesis of X₃₇ hold for every finite interval (c, d). Let $\alpha(x, s)$ be monotone increasing in s for all x, let ϕ be non-negative, and let $\gamma(x)$ be monotone increasing. Then if either of the integrals*

$$\int_{-\infty}^{\infty} \left[\int_a^b \phi(s)\,d_s\alpha(x, s)\right] d\gamma(x), \quad \int_a^b \phi(s)\,d_s \int_{-\infty}^{\infty} \alpha(x, s)\,d\gamma(x)$$

exists as a finite number, the other will exist and will have the same value.

---

* For X₃₇ and its predecessors, see Bray 1, Daniell 1.

Another theorem of a similar nature is

X_{39}. *Let $\{f_n(x)\}$ be a monotone increasing sequence of continuous functions, and let $g(x)$ be a monotone function. Let $\int_{-\infty}^{\infty} f_n(x) \, dg(x)$ exist for every n, and let*

$$\lim_{n \to \infty} \int_{-\infty}^{\infty} f_n(x) \, dg(x) = I < \infty.$$

Let $f(x) = \lim_{n \to \infty} f_n(x)$ exist everywhere and be continuous. Then[*]

$$I = \int_{-\infty}^{\infty} f(x) \, dg(x).$$

§ 3. The Riesz-Fischer Theorem.

We now come to the most important of the preliminary theorems of the book: the famous theorem of F. Riesz and E. Fischer[†]. In its original form, translated into the complex, it asserts that if $\sum_{-\infty}^{\infty} |a_n^2|$ converges, then there exists a function $f(x)$, defined over $(-\pi, \pi)$, and belonging to L_2, such that

$$\lim_{n \to \infty} \int_{-\pi}^{\pi} \left| f(x) - \sum_{-n}^{n} a_k e^{ikx} \right|^2 dx = 0.$$

However, this association of the theorem with a development, in terms of a particular set of functions, or even with any development whatever, is not essential. The gist of the theorem lies in the form given to it by H. Weyl[‡]. This asserts that:

X_{40}. *Given a sequence $\{f_n(x)\}$ of functions of L_2 on an interval (a, b) (or $(-\infty, \infty)$), for which*

(3·01) $$\lim_{m, n \to \infty} \int_a^b |f_m(x) - f_n(x)|^2 \, dx = 0,$$

there exists a function $f(x)$ belonging to L_2, such that[§]

(3·02) $$\lim_{n \to \infty} \int_a^b |f_n(x) - f(x)|^2 \, dx = 0.$$

---
[*] Cf. Daniell 1, p. 289, 7 (6).   [†] Cf. Riesz 1; Fischer 1.
[‡] Weyl 1; Plancherel 1 (p. 292).
[§] The converse transition from (3·02) to (3·01) follows directly from the Minkowski inequality.

Before we turn to the proof of this theorem, certain comments are perhaps appropriate. The first is that $\left(\int_a^b |f(x)|^2\right)^{\frac{1}{2}}$ is a "length" of the function $f(x)$ analogous to the length

$$\mathbf{x}\cdot\mathbf{x} = \left(\sum_1^3 x_{(k)}^2\right)^{\frac{1}{2}}$$

of the vector $\mathbf{x}$ with components $x(1), x(2), x(3)$. The analogue of (3·01) in the case of a sequence $\mathbf{x}_k$ of real vectors would be

$$\lim_{m,n\to\infty} (\mathbf{x}_m - \mathbf{x}_n)\cdot(\mathbf{x}_m - \mathbf{x}_n) = 0.$$

This would however imply that each component of $\mathbf{x}_m - \mathbf{x}_n$ must tend to 0, and by a well-known theorem of Cauchy, this is impossible unless there exists a number to which each component of $\mathbf{x}_m$ tends. Let the three numbers in question be $x(1), x(2), x(3)$. They will be components of a vector $\mathbf{x}$, and it will be easy to show that

$$\lim_{n\to\infty} (\mathbf{x}_n - \mathbf{x})\cdot(\mathbf{x}_n - \mathbf{x}) = 0.$$

Thus it is very natural to expect the Riesz-Fischer theorem to be valid, though as a function has an infinite set of components—namely, its values for each of a continuous set of arguments—it is too much to expect to prove it by a direct recourse to the Cauchy criterion of convergence.

Formula (3·03) represents a situation which will occur over a finite or infinite interval at many places in this book. We shall say that $f_n(x)$ *converges in the mean* to $f(x)$, which is its *limit in the mean* over (a, b). We shall write

(3·03) $$f(x) = \operatorname*{l.i.m.}_{n\to\infty} f_n(x),$$

with a similar notation where the parameter n tending to infinity is replaced by a variable tending (continuously or discretely) to some other limit.

It is important to realize that (3·03) does not imply that $f_n(x)$ converges at any point, and is not implied by the con-

INTRODUCTION

vergence of $f_n(x)$ at every point. As to the first, let $a = 0$, $b = 1$, and $k = [\log_2 n]$. Let

$$\begin{aligned} f_n(x) &= 0 \quad (x < (n-2^k)/2^k); \\ &= 1 \quad ((n-2^k)/2^k \leqslant x \leqslant (n-2^k)/2^k + 1); \\ &= 0 \quad ((n-2^k)/2^k + 1 < x). \end{aligned}$$

Clearly, $\int_0^1 |f_n(x)|^2 dx = 2^{-k} \leqslant 2/n$.

Thus $\text{l.i.m.}_{n \to \infty} f_n(x) = 0$. On the other hand, whatever quantity on $(0, 1)$ x may be, there is at least one f_n for $2^k < n < 2^{k+1}$ for which $f_n(x) = 1$, and at least one if $k > 1$ such that $f_n(x) = 0$. Thus for no x does $\lim_{n \to \infty} f_n(x)$ exist.

On the other hand, let $f_n(x) = n^{\frac{3}{2}} x e^{-n^2 x^2}$, and let $a = -1$, $b = 1$. It is easy to show that $f_n(x) \to 0$ for every x, while $\int_{-1}^1 |f_n(x)|^2 dx$ tends to $\pi^{\frac{1}{2}}/2$, and $f_n(x)$ does not converge in the mean. Thus convergence everywhere does not imply convergence in the mean. We shall later see, however, that if a sequence of functions $f_n(x)$ converges to a limit $f(x)$ almost everywhere on (a, b), and at the same time converges in the mean to a limit $g(x)$, then $f(x) = g(x)$ almost everywhere.

Now as to the proof of the Riesz-Fischer theorem: it is clear that if

$$\int_a^b |f_m(x) - f_n(x)|^2 dx < \epsilon^3,$$

then, except over a set S of measure not exceeding ϵ,

$$|f_m(x) - f_n(x)| < \epsilon.$$

Let $\sum_1^\infty \epsilon_\nu$ be a convergent sequence, and let n_k be so large that for $m > n_k$,

$$\int_a^b |f_{n_k}(x) - f_m(x)|^2 dx < \epsilon_\nu^3.$$

A consideration of (3·01) will show that this is always possible. We can choose our numbers n_k in such a way that $n_{k+1} > n_k$. Let the set of points on (a, b) for which $|f_{n_{k+1}}(x) - f_{n_k}(x)| \geqslant \epsilon_k$

be S_k. Let T_k be the logical sum $S_k + S_{k+1} + \ldots + S_{k+n} + \ldots$. Then except on T_k, which will have a measure not exceeding $\sum_{k}^{\infty} \epsilon_\nu$, we shall have

$$(3\cdot04) \qquad |f_{n_l}(x) - f_{n_m}(x)| < \sum_{k}^{\infty} \epsilon_\nu$$

for all l and m not less than k. Since, however,

$$\lim_{k \to \infty} \sum_{k}^{\infty} \epsilon_\nu = 0,$$

we see that by Cauchy's test, $f_{n_k}(x)$ converges to a limit $f(x)$ uniformly on the complementary set of any T_k, and hence it converges except over the set T common to all sets T_k. This is a null set by $\aleph_3$.

The function $f_A(x)$ is the bounded limit of the summable functions $(f_{n_k})_A(x)$, and is hence summable. Moreover, by bounded convergence,

$$\int_a^b |f_A(x) - (f_{n_k})_A(x)|^2 dx = \lim_{j \to \infty} \int_a^b |(f_{n_j})_A(x) - (f_{n_k})_A(x)|^2 dx$$
$$\leqslant 2A \sum_{k}^{\infty} \epsilon_\nu + (b-a) \left[\sum_{k}^{\infty} \epsilon_\nu \right]^2,$$

since (3·04) holds except on T_k, and

$$|(f_{n_j})_A(x) - (f_{n_k})_A(x)| \leqslant 2A.$$

Thus

$$(3\cdot05) \qquad \lim_{k \to \infty} \int_a^b |f_A(x) - (f_{n_k})_A(x)|^2 dx = 0,$$

and in view of (2·03) and the Minkowski inequality, (3·01) gives us

$$(3\cdot06) \qquad \lim_{n \to \infty} \int_a^b |f_A(x) - (f_n)_A(x)|^2 dx = 0,$$

since

$$\lim_{m, k \to \infty} \int_a^b |f_m(x) - f_{n_k}(x)|^2 dx = 0,$$

and hence

$$\lim_{m, k \to \infty} \int_a^b |(f_m)_A(x) - (f_{n_k})_A(x)|^2 dx = 0.$$

INTRODUCTION

By (3·05) and the Minkowski inequality,

$$(3\cdot 07) \quad \lim_{n \to \infty} \left[\int_a^b |f_A(x)|^2 \, dx - \int_a^b |(f_n)_A(x)|^2 \, dx \right] = 0.$$

Hence

$$\int_a^b |f_A(x)|^2 \, dx \leqslant \varlimsup_{n \to \infty} \int_a^b |(f_n)_A(x)|^2 \, dx \leqslant \varlimsup_{n \to \infty} \int_a^b |f_n(x)|^2 \, dx.$$

However, as in (3·07),

$$(3\cdot 071) \quad \lim_{m,\, n \to \infty} \left[\int_a^b |f_m(x)|^2 \, dx - \int_a^b |f_n(x)|^2 \, dx \right] = 0,$$

and by Cauchy's criterion,

$$(3\cdot 072) \quad \lim_{n \to \infty} \int_a^b |f_n(x)|^2 \, dx = L$$

exists and is finite. Thus

$$(3\cdot 08) \quad \int_a^b |f_A(x)|^2 \, dx \leqslant L,$$

and as $|f_A(x)|^2$ is an increasing function of A,

$$\lim_{A \to \infty} \int_a^b |f_A(x)|^2 \, dx = I \leqslant L$$

exists. Thus by X_{11}, since $|f_A(x)|^2$ converges monotonely to $|f(x)|^2$,

$$\int_a^b |f(x)|^2 \, dx = I \leqslant L,$$

and $f(x)$ belongs to L_2. Hence

$$\int_a^b |f(x) - f_A(x)|^2 \, dx$$

exists, and, again by X_{11},

$$\lim_{A \to \infty} \int_a^b |f(x) - f_A(x)|^2 \, dx = 0.$$

Thus by (3·06) we may choose A so large, and then n so large, that by the Minkowski inequality,

$$(3\cdot 09) \quad \int_a^b |f(x) - (f_n)_A(x)|^2 \, dx < \epsilon.$$

On the other hand, let m be so large that, for $n > m$,
$$\int_a^b |f_m(x) - f_n(x)|^2 \, dx < \epsilon.$$
By (2·015) we shall have, for any A,
$$\int_a^b |(f_m)_A(x) - (f_n)_A(x)|^2 \, dx < \epsilon.$$
Let A be so large that
$$\int_a^b |f_m(x) - (f_m)_A(x)|^2 \, dx < \epsilon.$$
By the Minkowski inequality, if $n > m$,
$$\int_a^b |f_n(x) - (f_n)_A(x)|^2 \, dx < 9\epsilon,$$
and by (3·09) we may choose A so large, and then n so large that
$$(3·10) \qquad \int_a^b |f_n(x) - f(x)|^2 \, dx < 16\epsilon.$$
Here A does not appear. Formula (3·10) however immediately leads to (3·02), and the Riesz-Fischer theorem is established in the case where (a, b) is finite.

When the interval is $(-\infty, \infty)$, this argument establishes the existence of a function $f(x)$ belonging to L_2 over any finite interval, and the limit in the mean of $f_n(x)$ over any such interval. However, the argument of (3·071), (3·072) is valid over the infinite interval. Thus, over *any* finite interval,
$$\int_a^b |f(x)|^2 \, dx \leqslant \lim_{n \to \infty} \int_{-\infty}^{\infty} |f_n(x)|^2 \, dx,$$
from which it appears that $f(x)$ belongs to L_2 over $(-\infty, \infty)$. We have
$$\int_{-\infty}^{\infty} |_A\phi(x) - {}_A\psi(x)|^2 \, dx \leqslant \int_{-\infty}^{\infty} |\phi(x) - \psi(x)|^2 \, dx$$
whenever ϕ and ψ belong to L_2 over $(-\infty, \infty)$, and if we replace f_A and $(f_m)_A$ by $_A f$ and $_A(f_m)$ in the argument (3·08)—(3·10),

we shall establish the Riesz-Fischer theorem in the infinite case. It is clear that if ϕ belongs to L_2,

$$\int_{-\infty}^{\infty} |\phi(x) - \phi_A(x)|^2 \, dx \to 0,$$

as the integral is merely

$$\int_{-\infty}^{\infty} |\phi(x)|^2 \, dx - \int_{-A}^{A} |\phi(x)|^2 \, dx,$$

which tends to zero with increasing A by the definition of an infinite integral. The rest of the argument is entirely unaltered.

We have thus proved the Riesz-Fischer theorem. Incidentally we have shown that if $f(x) = \underset{n\to\infty}{\text{l.i.m.}} f_n(x)$, $f(x)$ is almost everywhere the limit of a subsequence of $f_n(x)$. This establishes:

X$_{41}$. *If*

$$\underset{n\to\infty}{\text{l.i.m.}} f(x) = f(x),$$

and almost everywhere

$$\lim_{n\to\infty} f_n(x) = g(x),$$

then $f(x) = g(x)$ almost everywhere.

Proposition X$_{41}$ clearly remains true for limits in the mean with a continuously varying parameter, and for other limits than ∞ for n. A direct corollary of the Schwarz inequality is:

X$_{41a}$. *If*

$$\underset{n\to\infty}{\text{l.i.m.}} f_n(x) = f(x)$$

over (a, b) and $g(x)$ belongs to L_2 over (a, b), then

$$\int_a^b f_n(x) g(x) \, dx \to \int f(x) g(x) \, dx.$$

For

$$\left[\int_a^b [f_n(x) - f(x)] g(x) \, dx\right]^2 \leq \int_a^b [f_n(x) - f(x)]^2 \, dx \int_a^b [g(x)]^2 \, dx$$

and consequently tends to 0.

Another theorem resulting from the proof of X$_{40}$ is:

X$_{42}$. *If the functions of the sequence $f_n(x)$ are all increasing functions of x, and*

$$\underset{}{\text{l.i.m.}} f_n(x) = f(x),$$

then $f(x)$ differs from a monotone increasing function at most over a null set. For except over such a set, $f(x)$ is the limit of a sequence of monotone functions, and is hence monotone increasing. It results that the values of $f(x)$ thus obtained have a greatest lower bound for arguments to the left of x and a least upper bound for arguments to the right of x. If we call these $\phi(x-0)$ and $\phi(x+0)$ respectively, and define $\phi(x)$ at any point as $\frac{1}{2}(\phi(x-0)+\phi(x+0))$, $\phi(x)$ will be monotone increasing, and will only differ from $f(x)$ over a null set.

§4. Developments in Orthogonal Functions.

A set of complex functions $\{\phi_n(x)\}$ of the real variable x is said to be *orthogonal* over the interval (a, b) or $(-\infty, \infty)$ if all the functions of the set belong to L_2, and

$$(4 \cdot 01) \qquad \int \phi_m(x) \overline{\phi_n(x)} \, dx = 0, \quad (m \neq n)$$

the integral being taken over the appropriate interval. [In this section we shall not indicate the interval of integration, which will be the same throughout, and shall even write for simplicity such an integral as (4·01) as $\int \phi_m \bar{\phi}_n$.] A function $\phi_n(x)$ is said to be *normal* if $\int |\phi_n|^2 = 1$, and if all the functions of an orthogonal set are normal, the set will be said to be normal. Clearly if $\{\phi_n(x)\}$ is normal, $\{\overline{\phi_n(x)}\}$ is also normal.

It will be remembered that we have shown that a segment of a line may be mapped upon a square or cube or indeed upon a hypercube of any finite number of dimensions, in such a way that one-dimensional measure is mapped into measure in the appropriate number of dimensions. It follows that a set S of finite measure in any number of dimensions may be mapped on a linear segment with preservation of measure. If $f(P)$ is a function defined for all points P of S, and x is the corresponding point of the segment $(0, 1)$ to P under the mapping in question, $f(P)$ will generate a function $F(x)$, with the property that $F(x)=f(P)$. Now let $\{\phi_n(P)\}$ be a normal set of functions over S, in the sense that

$$\iint_S \phi_m(P) \phi_n(P) \, dV = \begin{cases} 0 & [m \neq n] \\ 1 & [m = n] \end{cases},$$

INTRODUCTION 35

where dV is the element of volume in a space of the number of dimensions of S. If $\Phi_n(x)$ corresponds to $\phi_n(P)$ as $F(x)$ does to $f(P)$, we shall have

$$\int_0^1 \Phi_m(x)\,\Phi_n(x)\,dx = \begin{cases} 0 & [m \neq n] \\ 1 & [m = n] \end{cases}.$$

Thus the entire theory of normal sets in a space of any number of dimensions is a mere translation of the theory of normal sets over an interval, and the results of this section may be immediately applied without restriction as to the numbers of dimensions. There is no difficulty in extending the theory to the case where S is measurable in any finite region, but is not finite. The theory of normal sets of functions over such a region is equivalent to the theory of normal sets over $(-\infty, \infty)$, which again does not differ essentially from the theory of normal sets over a finite interval.

We return to the one-dimensional case. Let $f(x)$ belong to L_2, and let $\phi_1(x), \ldots, \phi_n(x)$ be a finite normal set. Then we wish to establish:

X_{43}. *The quadratic polynomial*

$$\int \left| f(x) - \sum_1^n a_k \phi_k(x) \right|^2 dx$$

assumes its minimum value—which will be a proper minimum—for

(4·02) $$a_k = \int f \bar{\phi}_k.$$

To see this, let us notice that

(4·03) $\int | f - \Sigma a_k \phi_k |^2 = \int (f - \Sigma a_k \phi_k)(\bar{f} - \Sigma \bar{a}_k \bar{\phi}_k)$

$= \int | f |^2 - \Sigma a_k \int \bar{f} \phi_k - \Sigma \bar{a}_k \int f \bar{\phi}_k + \Sigma a_k \bar{a}_k$

$= \int | f |^2 - \Sigma | \int f \bar{\phi}_k |^2 + \Sigma | a_k - \int f \bar{\phi}_k |^2.$

The last term is the only one in the last line which contains an a_k, and is non-negative, being a minimum only when it is zero and (4·02) holds.

We have incidentally established:

X_{44}. (4·04) $$\int | f^2 | \geq \sum_1^n \left| \int f \bar{\phi}_k \right|^2,$$

since (4·01) is positive, and since when (4·02) holds, it reduces to
$$\int |f|^2 - \Sigma \left| \int f \bar{\phi}_k \right|^2.$$

Now let $f(x)$ belong to L_2, and let $\{\phi_n(x)\}$ be a denumerable infinite normal set. It follows from X$_{44}$, applied to any finite subset, that

X$_{45}$.
$$\int |f|^2 \geq \sum_{1}^{\infty} \left| \int f \bar{\phi}_k \right|^2.$$

This is known as the Bessel inequality.

Now let $\{\phi_n(x)\}$ be a denumerable normal set, and let $\sum_{1}^{\infty} |a_n|^2$ converge. If we put
$$f_n(x) = \sum_{1}^{n} a_k \phi_k(x),$$
and $m > n$, we have
$$\int |f_m - f_n|^2 = \sum_{n+1}^{m} |a_k|^2.$$

We may verify this by carrying out the indicated integration, and remembering that $\int \phi_j \bar{\phi}_k$ is 0 or 1 according as j and k are distinct or identical. Hence
$$\lim_{m, n \to \infty} \int |f_m - f_n|^2 = 0,$$
and by X$_{40}$, we obtain:

X$_{46}$. *There exists a function $f(x)$ belonging to L_2, and such that*
$$\lim_{n \to \infty} \int |f_n - f|^2 = 0.$$

We have already introduced the notation
$$f(x) = \underset{n \to \infty}{\text{l.i.m.}} f_n(x),$$
but we shall also find the equivalent notation
$$f(x) \sim \sum_{1}^{\infty} a_k \phi_k(x)$$
convenient. Of course, this notation does not imply the convergence of the infinite sum in the ordinary sense.

We thus obtain:

X$_{47}$. *If $f(x)$ belongs to L_2, and $\{\phi_n\}$ is a denumerable normal set, then there exists a function*
$$g(x) \sim f(x) - \sum_{1}^{\infty} \phi_n(x) \int f(x) \, \overline{\phi_n(x)} \, dx$$

INTRODUCTION

belonging to L_2. Moreover,

(4·05) $$\int g(x)\,\overline{\phi_n(x)}\,dx = 0$$

for every n.

An immediate corollary is:

X_{48}. *If $\{\phi_n\}$ is a denumerable normal set, the following two statements are equivalent:* (1) *If $f(x)$ belongs to L_2,*

(4·055) $$f(x) \sim \sum_1^\infty \phi_n(x) \int f(x)\,\overline{\phi_n(x)}\,dx,$$

and (2) *there exists no function $g(x)$ not equivalent to 0 for which* (4·05) *is true.*

In case either of the propositions of X_{48} is valid, $\{\phi_n\}$ is said to be *closed* or *complete*. It follows from X_{43} that:

X_{49}. *The denumerable normal set $\{\phi_n\}$ is complete when and only when, whenever $f(x)$ belongs to L_2 and $\epsilon > 0$, there exists an integer N and a polynomial $\sum_1^N a_k \phi_k(x)$, such that*

(4·06) $$\int \left| f(x) - \sum_1^N a_k\,\phi_k(x) \right|^2 dx < \epsilon.$$

Here we have only to remember that by X_{43},

$$\int \left| f(x) - \sum_1^N \phi_k(x) \int f\,\bar{\phi}_k \right|^2 dx < \epsilon.$$

An even less restrictive condition for completeness is given by:

X_{50}. *The denumerable normal set $\{\phi_n\}$ is complete when and only when, whenever $f(x)$ is a step-function belonging to L_2 and $\epsilon > 0$, there exists an integer N and a polynomial $\sum a_k \phi_k(x)$ such that* (4·06) *holds.* This follows at once from X_{49}, X_{30}, and the Minkowski inequality.

By the definition of closure, X_{48}, and (4·03), we obtain the Hurwitz-Parseval theorem:

X_{51}. *If $\{\phi_n\}$ is a complete denumerable normal set, and f belongs to L_2, then*

$$\int |f(x)|^2\,dx = \sum_1^\infty \left| \int f\,\bar{\phi}_n \right|^2.$$

It immediately results that if $f(x)$ and $g(x)$ belong to L_2,

$$\begin{cases} (A) & \int |f(x)+g(x)|^2\,dx = \sum_1^\infty \left|\int f\bar{\phi}_n + \int g\bar{\phi}_n\right|^2; \\ (B) & \int |f(x)-g(x)|^2\,dx = \sum_1^\infty \left|\int f\bar{\phi}_n - \int g\bar{\phi}_n\right|^2; \\ (C) & \int |f(x)+ig(x)|^2\,dx = \sum_1^\infty \left|\int f\bar{\phi}_n + i\int g\bar{\phi}_n\right|^2; \\ (D) & \int |f(x)-ig(x)|^2\,dx = \sum_1^\infty \left|\int f\bar{\phi}_n - i\int g\bar{\phi}_n\right|^2. \end{cases}$$

If we now perform the symbolic multiplication
$$[(A)-(B)+i(C)-i(D)]/4,$$
we obtain

(4·07) $$\int f(x)\overline{g(x)}\,dx = \sum_1^\infty \int f\bar{\phi}_n \int \bar{g}\phi_n.$$

The Parseval theorem asserts:

X_{52}. *If $\{\phi_n\}$ is a complete denumerable normal set, and f and g belong to L_2, (4·07) holds.*

There is an interesting process known as *normalization*, by which it is possible to obtain a normal set of functions out of every finite or denumerable set of functions of L_2. Let $\{\psi_n(x)\}$ be such a set, all having the same finite or infinite interval of definition. Let us consider $\psi_1(x)$. If it is equivalent to 0, we discard it; otherwise we put

(4·08) $$\phi_1(x) = \psi_1(x)/(\int|\psi_1|^2)^{\frac{1}{2}}.$$

If we have discarded $\psi_1(x)$, we proceed until we reach a function ψ_n not equivalent to 0, and replace ψ_1 by ψ_n in (4·08). Having obtained ϕ_1, we proceed along the ψ_k's until we find the first one not equivalent to any function of the form $a\phi_1$. If this is, for example, $\psi_2(x)$, we put

$$\phi_2(x) = (\psi_2(x) - \phi_1(x)\int\psi_2\bar{\phi}_1)/\{\int|\psi_2|^2 - |\int\psi_2\bar{\phi}_1|^2\}^{\frac{1}{2}}.$$

We now eliminate all functions equivalent to some $a\phi_1 + b\phi_2$, and if, say, ψ_3 is the first remaining function, we put

$$\phi_3(x) = \{(\psi_3(x) - \phi_1(x)\int\psi_3\bar{\phi}_1 - \phi_2(x)\int\psi_3\bar{\phi}_2)/\{\int|\psi_3|^2$$
$$- |\int\psi_3\bar{\phi}_1|^2 - |\int\psi_3\bar{\phi}_2|^2\},$$

and so on. It may be verified by direct integration that the set $\{\phi_n\}$ is normal, and that every function which can be approximated in the mean L_2 by polynomials in $\{\psi_n\}$ may be so approximated by polynomials in $\{\phi_n\}$, and vice versa, since every polynomial in the one is in fact equivalent to a polynomial in the other. Combining this fact with X_{47}, X_{48}, and X_{49}, we obtain:

X_{53}. *If $\{\psi_n\}$ is any denumerable set of L_2, the following two propositions are equivalent:* (1) *There exists no function $g(x)$ of L_2, such that*

$$\int g(x)\overline{\psi_n(x)}\,dx = 0$$

for every n; (2) *If $f(x)$ belongs to L_2 and $\epsilon > 0$, there exists an integer N and a polynomial*

$$\sum_1^N a_k \psi_k(x),$$

such that $\quad\displaystyle\int \left| f(x) - \sum_1^N a_k \psi_k(x) \right|^2 dx < \epsilon.$

If either of these statements is true, we shall say that $\{\psi_k\}$ is complete or closed.

Many interesting normal sets of functions are obtained by the process of normalization applied to sequences of elementary functions. Cases in point are:

(1) $\psi_n(x) = x^n$ $(n = 0, 1, \ldots)$. Interval of integration $(-1, 1)$. The $\phi_n(x)$ are (apart from constant factors) the Legendre polynomials $P_n(x)$.

(2) $\psi_n(x) = x^n e^{-x^2/2}$ $(n = 0, 1, \ldots)$. Interval of integration $(-\infty, \infty)$. The $\phi_n(x)$ are (apart from constant factors) the Hermite functions $H_n(x) e^{-x^2/2}$, where the functions $H_n(x)$ are the Hermite polynomials.

(3) $\psi_n(x) = x^n e^{-x}$ $(n = 0, 1, \ldots)$. Interval of integration $(0, \infty)$. The $\phi_n(x)$ are (apart from constant factors) the Laguerre functions $L_n(x) e^{-x}$, where the functions $L_n(x)$ are the Laguerre polynomials.

The exercise of computing the first few functions of each of these normal sequences is left to the reader.

INTRODUCTION

A particularly important normal sequence is the sequence

$$\frac{1}{\sqrt{2\pi}} e^{inx} \qquad (n = \ldots, -2, -1, 0, 1, 2, \ldots),$$

where the interval of integration is $(-\pi, \pi)$. The fact that the quantity n has a range extending to infinity in both directions introduces no real complication. The formal equivalent of (4·055) is now

$$(4\cdot09) \qquad f(x) \sim \sum_{-\infty}^{\infty} \frac{1}{2\pi} e^{inx} \int_{-\pi}^{\pi} f(y) e^{-iny} \, dy,$$

which is to be interpreted as meaning that

$$\underset{n\to\infty}{\text{l.i.m.}} \left(f(x) - \sum_{-n}^{n} \frac{1}{2\pi} e^{inx} \int_{-\pi}^{\pi} f(y) e^{-iny} \, dy \right) = 0.$$

We wish to prove that (4·09) is true for every function $f(x)$ belonging to L_2. The series of (4·09) is known as the Fourier series of $f(x)$, and we wish to prove that the Fourier series of any function of L_2 not merely converges in the mean, but converges in the mean to *that particular function*.

To this end, let us consider the series

$$(4\cdot10) \qquad \sum_{-\infty}^{\infty} \frac{r^{|n|}}{2\pi} e^{inx} \int_{-\pi}^{\pi} f(y) e^{-iny} \, dy.$$

This series converges uniformly in x and absolutely for any r such that $|r| < 1$. To see this, let us notice that, if $m > n > 0$,

$$\left\{ \left[\sum_{n+1}^{m} + \sum_{-m}^{1-n} \right] \left| \frac{r^{|k|}}{2\pi} e^{ikx} \int_{-\pi}^{\pi} f(y) e^{-iky} \, dy \right| \right\}^2$$

$$\leq \left[\sum_{n+1}^{m} + \sum_{-m}^{1-n} \right] \left| \int_{-\pi}^{\pi} f(y) e^{-iky} \, dy \right|^2 \left[\sum_{n+1}^{m} + \sum_{-m}^{1-n} \right] \frac{r^{2|k|}}{2\pi}$$

$$\leq \int_{-\pi}^{\pi} |f(y)|^2 \, dy \, \frac{2r^{|n|}}{1 - r^{|n|}}$$

in view of the appropriate form of the Schwarz inequality and of X_{44}. Thus

$$\overline{\lim_{m,n\to\infty}} \left\{ \sum_{-m}^{m} \left| \frac{r^{|k|}}{2\pi} e^{ikx} \int_{-\pi}^{\pi} f(y) e^{-iky} \, dy \right| \right.$$

$$\left. - \sum_{-n}^{n} \left| \frac{r^{|k|}}{2\pi} e^{ikx} \int_{-\pi}^{\pi} f(y) e^{-iky} \, dy \right| \right\}$$

$$\leq \int_{-\pi}^{\pi} |f(y)|^2 \, dy \, \overline{\lim_{m,n\to\infty}} \left\{ \frac{2r^{|m|}}{1 - r^{|m|}} + \frac{2r^{|n|}}{1 - r^{|n|}} \right\} = 0,$$

INTRODUCTION

and the absolute and uniform convergence of (4·10) is demonstrated.

Formally, we may write (4·10) as

$$\frac{1}{2\pi}\int_{-\pi}^{\pi} f(y) \sum_{-\infty}^{\infty} r^{|n|} e^{in(x-y)} dy.$$

The series
$$\sum_{-\infty}^{\infty} r^{|n|} e^{in(x-y)}$$

converges absolutely and uniformly for $|r|<1$, since it is a geometric progression, and is equal to

$$1 + \sum_{1}^{\infty} r^n e^{in(x-y)} + \sum_{1}^{\infty} r^n e^{-in(x-y)}$$
$$= 1 + \frac{re^{i(x-y)}}{1-re^{i(x-y)}} + \frac{re^{-i(x-y)}}{1-re^{-i(x-y)}}$$
$$= \frac{1-r^2}{1-2r\cos(x-y)+r^2}.$$

We have
$$\left|\frac{1-r^2}{1-2r\cos(x-y)+r^2}\right| \leq \frac{1-r}{1+r}.$$

Thus, by bounded convergence, we have:

X_{54}. *If $f(x)$ belongs to L_1, and hence if $f(x)$ belongs to L_2,*

(4·11) $\quad \sum_{-\infty}^{\infty} \frac{r^{|n|}}{2\pi} e^{inx} \int_{-\pi}^{\pi} f(y) e^{-iny} dy$

$$= \frac{1}{2\pi}\int_{-\pi}^{\pi} f(y) \frac{1-r^2}{1-2r\cos(x-y)+r^2} dy.$$

We now shall introduce as a lemma:

X_{55}. *Let $K(x, y, r)$ be non-negative. Let*

(4·12) $\quad \lim_{r \to 1-0} \int_a^b K(x, y, r) dy = 1$

uniformly in y for $a < x < b$, and if $\epsilon > 0$, let

(4·13) $\quad \lim_{r \to 1} K(x, y, r) = 0$

uniformly for all x and y in (a, b) for which $|y-x| > \epsilon$. Let $f(x)$ be bounded over (a, b). Then

(4·14) $\quad \lim_{r \to 1} \int_a^b K(x, y, r) f(y) dy = f(x)$

boundedly over the set of points interior to (a, b) at which $f(x)$ is continuous. This remains true if we replace (a, b) by $(-\infty, \infty)$.

To prove this, let us put $a + \epsilon < x < b - \epsilon$, and let us note that

(4·15) $$\left| f(x) - \int_a^b K(x, y, r) f(y)\, dy \right|,$$

by (4·12), is asymptotically equal to

(4·16) $$\left| \int_a^b K(x, y, r)(f(x) - f(y))\, dy \right|.$$

This again does not exceed

(4·17) $$2 \limsup |f(x)| \left[\int_a^{x-\epsilon} + \int_{x+\epsilon}^b \right] K(x, y, r)\, dy$$
$$+ \lim_{|y-x|<\epsilon} \sup |f(x) - f(y)| \int_a^b K(x, y, r)\, dy.$$

At any point of continuity of $f(x)$, we may choose ϵ so small that the second term of (4·17) is less than η in view of (4·12). We may also by (4·13) choose r so near 1 that the first term of (4·17) is less than η. It follows that for r near enough to 1, (4·15) does not exceed 2η, and hence that it tends to 0 as $r \to 1$. The boundedness of (4·15) results from the fact that (4·16) does not exceed

$$2 \limsup |f(x)| \int_a^b K(x, y, r)\, dy.$$

The argument of X_{55} has further proved:

X_{56}. *On the hypothesis of X_{55}, (4·14) holds uniformly over any interval interior to (a, b) and to one in which $f(x)$ is continuous.*

We merely need bring in the fact that a function is uniformly continuous over any closed interval of continuity.

Another theorem which we shall use is:

X_{57}. *If we put*
$$a = -\pi, \quad b = \pi, \quad K(x, y, r) = \frac{1}{2\pi} \frac{1 - r^2}{1 - 2r \cos(x - y) + r^2},$$

the hypothesis of X_{55} will be satisfied, as far as it concerns K.

INTRODUCTION 43

Here

$$\frac{1}{2\pi}\int_{-\pi}^{\pi}\frac{1-r^2}{1-2r\cos(x-y)+r^2}dy = \frac{1}{2\pi}\int_{-\pi}^{\pi}\left(\sum_{-\infty}^{\infty} r^{|n|} e^{in(x-y)}\right)dy$$

$$= \frac{1}{2\pi}\int_{-\pi}^{\pi} dy = 1.$$

Thus (4·12) is established. Again, if $|x-y| > \epsilon$,

$$K(x, y, r) < \frac{1}{2\pi}\frac{1-r^2}{1-2r\cos\epsilon+r^2} < \frac{1-r^2}{2\pi\sin^2\epsilon},$$

since the minimum of $1 - 2r\cos\epsilon + r^2$ is attained for $r = \cos\epsilon$ and is

$$1 - 2\cos^2\epsilon + \cos^2\epsilon = \sin^2\epsilon.$$

The validity of (4·13) is thus assured. The positiveness of K is also clear from the same consideration, and X_{57} is established.

Now let $f(x)$ be a step-function. Clearly, by X_{57}, (4·14) tends boundedly to $f(x)$ as $r \to 1$ except at a finite set of points. Thus

$$\left| f(x) - \sum_{-\infty}^{\infty} \frac{r^{|n|}}{2\pi} e^{inx} \int_{-\pi}^{\pi} f(y) e^{-iny} dy \right|^2$$

tends boundedly to 0, and, by X_{12},

$$f(x) = \underset{r \to 1-0}{\text{l.i.m.}} \sum_{-\infty}^{\infty} \frac{r^{|n|}}{2\pi} e^{inx} \int_{-\pi}^{\pi} f(y) e^{-iny} dy.$$

Since the series here is uniformly convergent, it converges in the mean, and given ϵ, by the Minkowski inequality, we may find an r so near 1 and an N so large that

$$\int_{-\pi}^{\pi} \left| f(x) - \sum_{-N}^{N} \frac{r^{|n|}}{2\pi} e^{inx} \int_{-\pi}^{\pi} f(y) e^{-iny} dy \right|^2 dx < \epsilon.$$

Thus, by X_{50}, we may establish:

X_{58}. *The set of functions* $\dfrac{1}{\sqrt{2\pi}} e^{inx}$ *is complete.*

The integral (4·11) is known as the Poisson integral, and theorems X_{55}, X_{56}, X_{57} constitute the adaptation to this integral of Fejér's method for the discussion of Fourier series. Fejér's original method differs only in the choice of the function $K(x, y, r)$.

Let it be noted that, with the aid of X_{58}, we can prove:

X_{59}. *No normal set is more than denumerable.* Let us take our interval of integration to be $(-\pi, \pi)$. Let Σ be a non-denumerable normal set of functions over this interval. Let us consider the developments of the functions $\frac{1}{\sqrt{2\pi}} e^{inx}$ in functions of Σ. By X_{44}, each function $\frac{1}{\sqrt{2\pi}} e^{inx}$ is orthogonal to all but a denumerable set of functions of Σ, since otherwise there would be more than a finite set of quantities

$$\frac{1}{\sqrt{2\pi}} \int_{-\pi}^{\pi} e^{-inx} \phi_k(x)\, dx$$

for which ϕ_k belongs to Σ, all exceeding some ϵ in modulus. This contradicts (4·04). Thus, since a denumerable set of denumerable sets contains only a denumerable set of elements, all but a denumerable set of functions of Σ are orthogonal to *every* $\frac{1}{\sqrt{2\pi}} e^{inx}$, and hence by X_{58} are equivalent to 0. By definition, however, no normal set can contain a function equivalent to 0.

It is perhaps worth while to state the Hurwitz-Parseval and the Parseval theorems explicitly in the form appropriate for Fourier series. They are respectively:

X_{60}. *If $f(x)$ belongs to L_2,*

$$\int_{-\pi}^{\pi} |f(x)|^2\, dx = \frac{1}{2\pi} \sum_{-\infty}^{\infty} \left| \int_{-\pi}^{\pi} f(x) e^{inx}\, dx \right|^2;$$

and

X_{61}. *If $f(x)$ and $g(x)$ belong to L_2,*

$$\int_{-\pi}^{\pi} f(x)\, \bar{g}(x)\, dx = \frac{1}{2\pi} \sum_{-\infty}^{\infty} \int_{-\pi}^{\pi} f(x) e^{inx}\, dx \int_{-\pi}^{\pi} \bar{g}(x) e^{-inx}\, dx,$$

and hence

$$\int_{-\pi}^{\pi} f(x)\, g(x)\, dx = \frac{1}{2\pi} \sum_{-\infty}^{\infty} \int_{-\pi}^{\pi} f(x) e^{inx}\, dx \int_{-\pi}^{\pi} g(x) e^{-inx}\, dx.$$

INTRODUCTION

In particular, if g is furthermore of period 2π,

(4·18)
$$\int_{-\pi}^{\pi} f(x)g(y-x)\,dx = \frac{1}{2\pi}\sum_{-\infty}^{\infty}\int_{-\pi}^{\pi} f(x)e^{inx}\,dx \int_{-\pi}^{\pi} g(y-x)e^{-inx}\,dx$$
$$= \frac{1}{2\pi}\sum_{-\infty}^{\infty} e^{-iny}\int_{-\pi}^{\pi} f(x)e^{inx}\,dx \int_{-\pi}^{\pi} g(x)e^{inx}\,dx,$$

and

(4·19)
$$\int_{-\pi}^{\pi} f(x)g(x)e^{imx}\,dx = \frac{1}{2\pi}\sum_{-\infty}^{\infty}\int_{-\pi}^{\pi} f(x)e^{inx}\,dx \int_{-\pi}^{\pi} g(x)e^{i(m-n)x}\,dx.$$

The quantity $\dfrac{1}{2\pi}\displaystyle\int_{-\pi}^{\pi} f(x)g(y-x)\,dx$ is known as the *Faltung* of $f(x)$ and $g(x)$ (there is no good English word), and the sequence $\left\{\displaystyle\sum_{-\infty}^{\infty} a_n b_{m-n}\right\}$ as the Faltung of the sequences $\{a_n\}$ and $\{b_n\}$. The formal interpretation to be given to (4·18) and (4·19) is that the Fourier coefficients of the Faltung of two functions are the products of the corresponding Fourier coefficients of the individual functions, and that the Faltung of the Fourier coefficients of two functions yields the Fourier coefficients of their product.

CHAPTER I

PLANCHEREL'S THEOREM

§ 5. The Formal Theory of the Fourier Transform.

Up to the present, we have been discussing the harmonic analysis of functions $f(x)$ with period 2π, or (alternatively) of sequences of coefficients a_n defined only for integral values of n. It is interesting to see what formally becomes of the Fourier series when the period is $2A$ rather than 2π. The complete normal set of trigonometrical functions now becomes

$$\frac{1}{\sqrt{2A}} e^{\frac{n\pi i x}{A}},$$

and the formal Fourier series for $f(x)$ is

$$\frac{1}{2A} \sum_{-\infty}^{\infty} e^{\frac{n\pi i x}{A}} \int_{-A}^{A} f(y) e^{-\frac{n\pi i y}{A}} dy.$$

We may write this

(5·01) $\begin{cases} g\left(\dfrac{n\pi}{A}\right) = \dfrac{1}{\sqrt{2\pi}} \displaystyle\int_{-A}^{A} f(x) e^{-\frac{n\pi i x}{A}} dx; \\ f(x) \sim \dfrac{1}{\sqrt{2\pi}} \displaystyle\sum_{-\infty}^{\infty} g\left(\dfrac{n\pi}{A}\right) e^{\frac{n\pi i x}{A}} \Delta\left(\dfrac{n\pi}{A}\right). \end{cases}$

This pair of formulas tends to the form

(5·02) $\begin{cases} g(u) \sim \dfrac{1}{\sqrt{2\pi}} \displaystyle\int_{-\infty}^{\infty} f(x) e^{-iux} dx \\ f(x) \sim \dfrac{1}{\sqrt{2\pi}} \displaystyle\int_{-\infty}^{\infty} g(u) e^{iux} du, \end{cases}$

as A becomes infinite. In the first formula of (5·01) the interval of integration becomes larger, and in the second the

PLANCHEREL'S THEOREM

subdivision of the infinite interval of summation becomes finer and more nearly continuous. In the limit, the formulae (5·02) assume an almost perfect symmetry as between the function to be expanded and the coefficients of its development.

It will be noted that if we apply to $g(x)$ the same transformation which turned $f(x)$ into $g(x)$, we get $f(-x)$. A further application of the transformation yields $g(-x)$, and a fourth $f(x)$. Thus the transformation is of period 4.

The present chapter is devoted to the establishment of conditions for the validity of (5·02) in a certain extended sense. Before we go into the demonstration of this theorem, or even its enunciation, it is perhaps worth while to consider a number of concrete pairs such as (5·02), of *Fourier transforms*, as we shall call them. If, for example,

$$f(x) = 1 \quad (a < x < b); \quad = 0 \quad (a > x \text{ or } x > b),$$

we have $\quad g(u) = \dfrac{1}{\sqrt{2\pi}} \displaystyle\int_a^b e^{-iux}\, dx = \dfrac{e^{-iub} - e^{-iua}}{-iu\sqrt{2\pi}}.$

We shall then have

(5·03)
$$\frac{1}{\sqrt{2\pi}} \int_{-\infty}^{\infty} g(u) e^{iux}\, dx = -\frac{1}{2\pi i} \int_{-\infty}^{\infty} \frac{e^{iu(x-b)} - e^{iu(x-a)}}{u}\, du$$
$$= -\frac{1}{2\pi i} \int_{-\infty}^{\infty} \frac{\cos u(x-b) - \cos u(x-a)}{u}\, du$$
$$- \frac{1}{2\pi} \int_0^{\infty} \frac{\sin u(x-b) - \sin u(x-a)}{u}\, du.$$

Here the first integrand is odd, so that the integral vanishes. The second is even, so that (5·03) becomes

(5·04) $\quad \dfrac{1}{\pi} \displaystyle\int_0^{\infty} \dfrac{\sin u(x-a)}{u}\, du - \dfrac{1}{\pi} \int_0^{\infty} \dfrac{\sin u(x-b)}{u}\, du.$

Now let us remember that according as $A \gtreqless 0$,

$$\int_0^{\infty} \frac{\sin Au}{u}\, du = \int_0^{\pm\infty} \frac{\sin v}{v}\, dv,$$

where $v = Au$. Hence

$$\frac{1}{\sqrt{2\pi}} \int_{-\infty}^{\infty} g(u) e^{iux} dx = \frac{1}{\pi} \int_0^{\infty} \frac{\sin u}{u} du \, [\mathrm{sgn}\,(x-a) - \mathrm{sgn}\,(x-b)],$$

where sgn is used to indicate 0 if $x = 0$, and otherwise $x/|x|$. The integral $\int_0^{\infty} \frac{\sin u}{u} du$ is represented geometrically by a series of arches alternating in sign and decreasing to zero in area, and hence converges.

We have

$$\sum_{-n}^{n} e^{iku} = e^{-inu} \sum_0^{2n} e^{iku} = e^{-inu} \frac{e^{(2n+1)iu} - 1}{e^{iu} - 1}$$

$$= \frac{\sin \frac{2n+1}{2} u}{\sin \frac{u}{2}}.$$

Hence

(5·05) $$\int_0^{\pi} \frac{\sin \frac{2n+1}{2} u}{\sin \frac{u}{2}} du = \pi.$$

By the Riemann-Lebesgue theorem X_{19}, since $2/u - \mathrm{cosec}\, u/2$ is bounded over $(0, \pi)$,

(5·06) $$\lim_{n \to \infty} \int_0^{\pi} \sin \frac{2n+1}{2} u \left(\frac{2}{u} - \mathrm{cosec}\, \frac{u}{2} \right) du = 0,$$

and by combining (5·05) and (5·06), we get

$$\lim_{n \to \infty} 2 \int_0^{\pi} \frac{\sin \frac{2n+1}{2} u}{u} du = \pi.$$

This may be written

$$\lim_{n \to \infty} \int_0^{\frac{2n+1}{2}\pi} \frac{\sin v}{v} dv = \frac{\pi}{2}$$

by a change of variable, and since the Cauchy integral in question is convergent,

$$\int_0^{\infty} \frac{\sin u}{u} du = \frac{\pi}{2}.$$

PLANCHEREL'S THEOREM

Hence (5·04) becomes

(5·07) $\quad \frac{1}{2}\left[\operatorname{sgn}(x-a)-\operatorname{sgn}(x-b)\right]$

$$= \begin{cases} 0 & \text{if } x < a; \\ \frac{1}{2} & \text{if } x = a; \\ 1 & \text{if } a < x < b; \\ \frac{1}{2} & \text{if } x = b; \\ 0 & \text{if } b < x \end{cases}$$

$= f(x)$ except at a and b.

Again, let

$$f(x) = 1 - \frac{|x|}{a} \quad (|x| < a); \quad = 0 \quad (|x| \geq a); \quad a > 0.$$

We have
$$g(u) = \frac{1}{\sqrt{2\pi}} \int_{-a}^{a} \left(1 - \frac{|x|}{a}\right) e^{-iux} dx$$

$$= \sqrt{\frac{2}{\pi}} \int_{0}^{a} \left(1 - \frac{x}{a}\right) \cos ux \, dx$$

$$= \sqrt{\frac{2}{\pi}} \frac{1}{u} \int_{0}^{a} \left(1 - \frac{x}{a}\right) d(\sin ux)$$

$$= \sqrt{\frac{2}{\pi}} \int_{0}^{a} \frac{\sin ux}{ua} dx$$

$$= \sqrt{\frac{2}{\pi}} \frac{1 - \cos ua}{u^2 a}.$$

We shall have
$$\frac{1}{\pi} \int_{-\infty}^{\infty} \frac{1 - \cos ua}{u^2 a} e^{iux} dx$$

$$= \frac{1}{\pi} \int_{-\infty}^{\infty} \frac{\cos ux (1 - \cos ua)}{u^2 a} du,$$

as we see by throwing out the odd part of the integrand. This is again

(5·08) $\quad \dfrac{1}{\pi} \int_{-\infty}^{\infty} \dfrac{\cos ux - \frac{1}{2}\cos u(x+a) - \frac{1}{2}\cos u(x-a)}{u^2 a} du$

$$= -\frac{2}{\pi} \int_{0}^{\infty} \frac{\cos ux - \frac{1}{2}\cos u(x+a) - \frac{1}{2}\cos u(x-a)}{a} d\left(\frac{1}{u}\right)$$

$$= \frac{2}{\pi} \int_{0}^{\infty} \frac{-x \sin ux + \dfrac{x+a}{2} \sin u(x+a) + \dfrac{x-a}{2} \sin u(x-a)}{au} du$$

by an integration by parts. If we apply (5·07), this becomes

$$\frac{1}{a}\left(-x\operatorname{sgn} x + \frac{x+a}{2}\operatorname{sgn}(x+a) + \frac{x-a}{2}\operatorname{sgn}(x-a)\right)$$

$$= \frac{1}{a}\left(\left|\frac{x+a}{2}\right| + \left|\frac{x-a}{2}\right| - |x|\right)$$

$$\begin{cases} = \dfrac{|x|-|x|}{a} & \text{if } a < |x| \\[4pt] = \dfrac{a-|x|}{a} & \text{if } |x| < a \\[4pt] = 0 & \text{if } |x| = a \end{cases}$$

$$= f(x).$$

Next, let $f(x) = e^{-x^2/2}$.

We shall have

(5·09) $$g(u) = \frac{1}{\sqrt{2\pi}} \int_{-\infty}^{\infty} e^{-x^2/2 - iux} dx$$

$$= \frac{1}{\sqrt{2\pi}} \int_{-\infty}^{\infty} e^{-(x+iu)^2/2} e^{-u^2/2} dx.$$

Inasmuch as $e^{-x^2/2}$ is entire [integral] and tends rapidly to zero uniformly for any finite range of $I(x)$ as $R(x) \to \pm\infty$, we may deform the path of integration in (5·09), so as to obtain

(5·10) $$g(u) = e^{-u^2/2} \frac{1}{\sqrt{2\pi}} \int_{-\infty}^{\infty} e^{-x^2/2} dx.$$

The definite integral in (5·10) may be computed by

$$\left[\frac{1}{\sqrt{2\pi}} \int_{-\infty}^{\infty} e^{-x^2/2} dx\right]^2 = \frac{1}{2\pi} \int_{-\infty}^{\infty} e^{-x^2/2} dx \int_{-\infty}^{\infty} e^{-y^2/2} dy$$

$$= \frac{1}{2\pi} \int_0^{2\pi} d\theta \int_0^{\infty} r e^{-r^2/2} dr = 1.$$

As the integral in question is manifestly positive, we obtain

$$g(u) = e^{-u^2/2} = f(u).$$

On applying (5·02) we get, in precisely the same manner,

$$e^{-x^2/2} = \frac{1}{\sqrt{2\pi}} \int_{-\infty}^{\infty} g(u) e^{iux} du.$$

If we integrate by parts in the first formula of (5·02), we get

$$g(u) \sim \lim_{A \to \infty} \frac{1}{\sqrt{2\pi}} \int_{-A}^{A} f(x) \, d\frac{e^{-iux}}{-iu}$$

$$\sim \lim_{A \to \infty} \left\{ \frac{1}{\sqrt{2\pi}} \left[\frac{f(x) e^{-iux}}{-iu} \right]_{x=-A}^{x=A} + \frac{1}{\sqrt{2\pi}} \int_{-A}^{A} \frac{f'(x) e^{-iux}}{iu} \, dx \right\}.$$

Accordingly, if $f(x) \to 0$ as $x \to \pm \infty$, and if $f(x)$ is the integral of its derivative,

$$iu\, g(u) \sim \frac{1}{\sqrt{2\pi}} \int_{-\infty}^{\infty} f'(x) e^{-iux} dx.$$

Again, if we assume that we may differentiate the second formula under the sign of integration, we obtain

$$f'(x) \sim \frac{1}{\sqrt{2\pi}} \int_{-\infty}^{\infty} iu\, g(u) e^{iux} du.$$

Thus—formally at least—the linear operation of multiplication by iu applied to $g(u)$ corresponds to the linear operation of differentiation applied to $f(x)$. Similarly, the linear operation of differentiation applied to $g(u)$ corresponds to the operation of multiplication by $-ix$ applied to $f(x)$. Thus the linear operation $\left(\frac{d^2}{du^2} - u^2\right)$ applied to $g(u)$ corresponds to the linear operation $\left(\frac{d^2}{dx^2} - x^2\right)$ applied to $f(x)$.

§ 6. Hermite Polynomials and Hermite Functions.

It became clear at the end of the last section that the differential operator $\frac{d^2}{dx^2} - x^2$ is its own Fourier transform, and that the solutions of the homogeneous linear second order differential equation

(6·01) $$\frac{d^2 w}{dx^2} - x^2 w = \lambda w$$

might be expected to play an important part in the theory of the Fourier transform. To solve this equation, let us put $w = e^{-x^2/2} W$. Then (6·01) becomes

(6·02) $$W'' - 2xW' = (\lambda + 1) W.$$

Let us look for polynomial solutions of the type
(6·03) $$W = a_0 + a_1 x + \ldots + a_n x^n \quad (a_n \neq 0).$$
Then (6·02) assumes the form
$$\sum_{2}^{n} k(k-1) a_k x^{k-2} = \sum_{0}^{n} (2k + \lambda + 1) a_k x^k,$$
from which we see that (6·03) holds when and only when
$$a_{k-2} = \frac{k(k-1)}{2k + \lambda - 3} a_k \quad (k = 2, 3, \ldots, n);$$
$$a_{n-1}(2n + \lambda - 1) = a_n(2n + \lambda + 1) = 0.$$
Thus since $a_n \neq 0$,
$$\lambda = -(2n+1); \quad a_{n-1} = 0,$$
and all coefficients vanish for indices not of the same parity as n. No coefficient of non-negative index of the same parity as n and less than n can vanish, as
$$a_{k-2} = \frac{k(k-1)}{2k - 2n - 4} a_k.$$
The function W of (6·03) is then
$$a_n \left(x^n - \frac{n(n-1)}{4} x^{n-2} + \frac{n(n-1)(n-2)(n-3)}{4 \cdot 8} x^{n-4} \right.$$
$$\left. - \frac{n(n-1)(n-2)(n-3)(n-4)}{4 \cdot 8 \cdot 12} x^{n-6} + \ldots \right),$$
in the sense that this is always a solution of
(6·04) $$W'' - 2xW' + 2nW = 0,$$
and is the only polynomial solution. Thus (6·01) only has a solution of the form $P(x) e^{-x^2/2}$ (P a polynomial) when $\lambda = -(2n+1)$, it always has a solution of this type when n is a non-negative integer, and this solution is except for a constant factor $H_n(x) e^{-x^2/2}$, where
(6·05) $$H_n(x) = 2^n \left(x^n - \frac{n(n-1)}{4} x^{n-2} \right.$$
$$\left. + \frac{n(n-1)(n-2)(n-3)}{4 \cdot 8} x^{n-4} - \ldots \right).$$
These functions $H_n(x) e^{-x^2/2}$ of course vanish exponentially at $\pm \infty$, as do all their derivatives.

PLANCHEREL'S THEOREM

We shall now show that if we write

(6·06) $$\phi_n(x) = H_n(x) e^{-x^2/2},$$

then, if $m \neq n$,

(6·07) $$\int_{-\infty}^{\infty} \phi_m(x) \phi_n(x) dx = 0.$$

For by (6·01)

(6·08) $$\begin{cases} \phi_m'' - x^2 \phi_m = -(2m+1) \phi_m; \\ \phi_n'' - x^2 \phi_n = -(2n+1) \phi_n. \end{cases}$$

If we multiply the first equation in (6·08) by ϕ_n and the second by ϕ_m, and subtract, we obtain

$$\phi_m'' \phi_n - \phi_n'' \phi_m = 2(n-m) \phi_m \phi_n.$$

Upon integrating by parts over $(-\infty, \infty)$, we get

$$\int_{-\infty}^{\infty} \phi_m(x) \phi_n(x) dx$$

$$= \frac{1}{2(n-m)} \int_{-\infty}^{\infty} [\phi_m''(x) \phi_n(x) - \phi_n''(x) \phi_m(x)] dx$$

$$= \frac{1}{2(n-m)} \left\{ \left[\phi_m'(x) \phi_n(x) - \phi_n'(x) \phi_m(x) \right]_{-\infty}^{\infty} - \int_{-\infty}^{\infty} [\phi_m'(x) \phi_n'(x) - \phi_n'(x) \phi_m'(x)] dx \right\}$$

$$= 0.$$

Thus the set of real functions $\{\phi_n(x)\}$ $[n = 0, 1, 2, \ldots]$ is orthogonal, and if we put

(6·09) $$\psi_n(x) = \phi_n(x) \Big/ \left\{ \int_{-\infty}^{\infty} [\phi_n(x)]^2 \right\}^{\frac{1}{2}},$$

the set $\{\psi_n(x)\}$ is normal.

The functions $\{\phi_n(x)\}$ or $\{\psi_n(x)\}$ are their own Fourier transforms, apart from a constant factor. This follows from the facts:

(a) That equation (6·01) is invariant under Fourier transformation;

(b) That equation (6·01) has for any λ essentially at most one solution of the form $P(x) e^{x^2/2}$, where x is a polynomial;

54   PLANCHEREL'S THEOREM

(c) That such a solution is of the form $\sum_{1}^{n} a_k x^k e^{-x^2/2}$;

(d) That the Fourier transform of $x^k e^{-x^2/2}$ is $\left(i\dfrac{d}{dx}\right)^k e^{-x^2/2}$;

(e) That $\left(i\dfrac{d}{dx}\right)^k e^{-x^2/2}$ is of the form $P_n(x) e^{-x^2/2}$, where $P_n(x)$ is a polynomial of the nth degree, as can be verified by mathematical induction.

Thus the Fourier transform of a solution $\psi_n(x)$ of (6·01) is of the form $Q(x) e^{-x^2/2}$, where $Q(x)$ is a polynomial, and since it satisfies (6·01), it is of the form $c\psi_n(x)$. The constant c can only be ± 1 or $\pm i$. This results from the fact that the Fourier transform is of period 4. Thus $c^4 \psi_n(x) = \psi_n(x)$. We shall verify this result in the next few sections, where we use the functions $\psi_n(x)$ to obtain a new and more general definition of the Fourier transform.

In this connection, let us note that we have established (5·02) for $e^{-x^2/2}$, and consequently for its product by a polynomial of any order, as may be verified by repeated differentiation under the integral sign.

It is manifest that the functions $\psi_n(x)$ are except for a possible difference in sign the functions we obtain by normalizing the sequence $x^n e^{-x^2/2}$ [$n = 0, 1, 2, \ldots$]. The functions $H_n(x)$ are also given by

(6·10) $$H_n(x) = (-1)^n e^{x^2} \frac{d^n e^{-x^2}}{dx^n}.$$

This polynomial satisfies (6·04), as

$$\frac{d}{dx}(-1)^n e^{x^2}\frac{d^n e^{-x^2}}{dx^n} = 2x(-1)^n e^{x^2}\frac{d^n e^{-x^2}}{dx^n} + (-1)^n e^{x^2}\frac{d^{n+1} e^{-x^2}}{dx^{n+1}},$$

or, if $H_n(x)$ is defined by (6·10),

(6·11) $\qquad H_n'(x) = 2x H_n(x) - H_{n+1}(x).$

Thus

(6·12) $H_{n+1}(x) = 2x H_n(x) - H_n'(x);$
$\qquad H_{n+1}'(x) = 2H_n(x) + 2x H_n'(x) - H_n''(x);$
$\qquad H_{n+1}''(x) = 4H_n'(x) + 2x H_n''(x) - H_n'''(x).$

PLANCHEREL'S THEOREM

Hence

(6·13) $H_{n+1}'' - 2xH_{n+1}' + 2(n+1)H_n$
$= -H_n''' + 4xH_n'' + (2 - 4x^2 - 2n)H_n' + 4nxH_n$
$= -\dfrac{d}{dx}(H_n'' - 2xH_n' + 2nH_n) + 2x(H_n'' - 2xH_n' + 2nH_n).$

Again, by (6·10),

(6·14) $\qquad H_0(x) = e^{x^2} \cdot e^{-x^2} = 1,$

so that

(6·15) $\qquad H_0''(x) - 2xH_0'(x) + 2(0)H_0(x) = 0.$

Combining (6·15) and (6·13), we obtain by mathematical induction

(6·16) $\qquad H_n''(x) - 2xH_n'(x) + 2nH_n(x) = 0,$

so that $H_n(x)$ can differ at most by a constant factor from the value given in (6·05). To see that (6·05) and (6·10) are completely in accord, let us notice that it follows by mathematical induction from (6·12) and (6·14) that the term of highest degree in $H_n(x)$ is $2^n x^n$, which is in accordance with (6·05).

We shall call the polynomials $H_n(x)$ the *Hermite polynomials*, and the functions $\psi_n(x)$ the *Hermite functions**.

§ 7. The Generating Function of the Hermite Functions.

In consequence of (6·10),

(7·01) $\qquad \phi_n(x) = (-1)^n e^{x^2/2} \dfrac{d^n e^{-x^2}}{dx^n}.$

Thus

(7·02) $\qquad e^{-x^2/2 + 2\lambda x - \lambda^2} = e^{x^2/2} e^{-(x-\lambda)^2}$
$= \sum\limits_0^\infty e^{x^2/2} \dfrac{\lambda^n}{n!} (-1)^n \dfrac{d^n e^{-x^2}}{dx^n}$
$= \sum\limits_0^\infty \dfrac{\lambda^n \phi_n(x)}{n!}$

over $(-\infty, \infty)$.

---

* C. Hermite, "Sur un nouveau développement en série des fonctions," *C.R.* 58, 93-100 and 266-273. Also *Œuvres*, 2, 293-312, Paris, 1908. See also "Sur quelques développements en série de fonctions de plusieurs variables," *C.R.* 60, 370-377, 432-440, 461-466, 512-518. Courant-Hilbert, *Methoden der mathematischen Physik*, I, pp. 76, 77, Berlin, 1924.

Again, by an integration by parts,

(7·03)
$$\int_{-\infty}^{\infty} e^{x^2}\left(\frac{d^{n+1}}{dx^{n+1}}e^{-x^2}\right)^2 dx = \int_{-\infty}^{\infty} e^{x^2}\left(\frac{d^{n+1}}{dx^{n+1}}e^{-x^2}\right) d\left(\frac{d^n}{dx^n}e^{-x^2}\right)$$
$$= -\int_{-\infty}^{\infty}\left(\frac{d^n}{dx^n}e^{-x^2}\right) d\left(e^{x^2}\frac{d^{n+1}}{dx^{n+1}}e^{-x^2}\right)$$
$$= -\int_{-\infty}^{\infty} 2xe^{x^2}\left(\frac{d^n}{dx^n}e^{-x^2}\right)\left(\frac{d^{n+1}}{dx^{n+1}}e^{-x^2}\right) dx$$
$$-\int_{-\infty}^{\infty} e^{x^2}\left(\frac{d^n}{dx^n}e^{-x^2}\right)\left(\frac{d^{n+2}}{dx^{n+2}}e^{-x^2}\right) dx.$$

By formulae (6·07), (6·10) and (7·01), we obtain

(7·04) $$\int_{-\infty}^{\infty} e^{x^2}\left(\frac{d^n}{dx^n}e^{-x^2}\right)\left(\frac{d^{n+2}}{dx^{n+2}}e^{-x^2}\right) dx = 0.$$

Furthermore, by (6·11),
$$H_n''(x) = 2H_n(x) + 2xH_n'(x) - H_{n+1}'(x).$$

In combination with (6·16), this yields us

(7·05) $$H_{n+1}'(x) = (2n+2)H_n(x).$$

On substituting $n+1$ for n in (6·11), we obtain
$$H_{n+1}'(x) = 2xH_{n+1}(x) - H_{n+2}(x).$$

Combining this with (7·05), we get
$$2xH_{n+1}(x) = H_{n+2}(x) + 2(n+1)H_n(x),$$

or, by (6·10) and (7·01),
$$2x\phi_{n+1}(x) = \phi_{n+2}(x) + 2(n+1)\phi_n(x).$$

If we substitute this value and remember (7·04), we get
$$\int_{-\infty}^{\infty} 2xe^{x^2}\left(\frac{d^n}{dx^n}e^{-x^2}\right)\left(\frac{d^{n+1}}{dx^{n+1}}e^{-x^2}\right) dx$$
$$= -2(n+1)\int_{-\infty}^{\infty} e^{x^2}\left(\frac{d^n}{dx^n}e^{-x^2}\right)^2 dx.$$

If we substitute this and (7·04) in (7·03), we obtain the recursion formula
$$\int_{-\infty}^{\infty} e^{x^2}\left(\frac{d^{n+1}}{dx^{n+1}}e^{-x^2}\right)^2 dx = 2(n+1)\int_{-\infty}^{\infty} e^{x^2}\left(\frac{d^n}{dx^n}e^{-x^2}\right)^2 dx,$$

PLANCHEREL'S THEOREM

or, by (7·01),

$$\int_{-\infty}^{\infty} [\phi_{n+1}(x)]^2 \, dx = 2(n+1) \int_{-\infty}^{\infty} [\phi_n(x)]^2 \, dx.$$

Since $\int_{-\infty}^{\infty} [\phi_0(x)]^2 \, dx = \int_{-\infty}^{\infty} e^{-x^2} \, dx = \sqrt{\pi},$

we have by mathematical induction

$$\int_{-\infty}^{\infty} [\phi_n(x)]^2 \, dx = 2^n \, n! \, \sqrt{\pi}.$$

By (6·09), $\psi_n(x) = \phi_n(x) / [2^n n! \sqrt{\pi}]^{\frac{1}{2}},$

and by (7·02),

$$e^{-x^2/2 + 2\lambda x - \lambda^2} = \sum_0^\infty \lambda^n \left[\frac{2^n \sqrt{\pi}}{n!} \right]^{\frac{1}{2}} \psi_n(x).$$

However, $\sum_0^\infty \lambda^{2n} \frac{2^n \sqrt{\pi}}{n!} < \infty$

by the ratio test, so that

$$\underset{N \to \infty}{\text{l.i.m.}} \sum_0^N \lambda^n \left[\frac{2^n \sqrt{\pi}}{n!} \right]^{\frac{1}{2}} \psi_n(x)$$

exists by the Riesz-Fischer theorem. As a limit in the mean of a series can differ from its limit at most over a null set, we have

$$e^{-x^2/2 + 2\lambda x - \lambda^2} \sim \sum_0^\infty \lambda^n \left[\frac{2^n \sqrt{\pi}}{n!} \right]^{\frac{1}{2}} \psi_n(x).$$

Formally, if

(7·06) $\quad K(x, y, t) = \sum_0^\infty t^n \psi_n(x) \psi_n(y),$

we should have

$$e^{-y^2/2 + 2\lambda t y - \lambda^2 t^2} \sim \sum_0^\infty \lambda^n t^n \left[\frac{2^n \sqrt{\pi}}{n!} \right]^{\frac{1}{2}} \psi_n(y)$$

$$\sim \sum_0^\infty \int_{-\infty}^{\infty} t^n \psi_n(y) [\psi_n(x)]^2 \, dx \lambda^n \left[\frac{2^n \sqrt{\pi}}{n!} \right]$$

$$= \int_{-\infty}^{\infty} K(x, y, t) \sum_0^\infty \lambda^n \left[\frac{2^n \sqrt{\pi}}{n!} \right]^{\frac{1}{2}} \psi_n(x) \, dx$$

$$= \int_{-\infty}^{\infty} K(x, y, t) \, e^{-x^2/2 + 2\lambda x - \lambda^2} \, dx.$$

That is,

(7·07)
$$\int_{-\infty}^{\infty} K(x, y, t) \exp\left\{\frac{y^2 - x^2}{2} + 2\lambda(x - yt) - \lambda^2(1 - t^2)\right\} dx = 1.$$

In solving (7·07), it is natural to seek for a function $K(x, y, t)$ in the form $D \exp(Ax^2 + Bxy + Cy^2)$, where A, B, C, and D depend on t. In the first place, we wish to make

$$-(A - \tfrac{1}{2})x^2 - Bxy - (C + \tfrac{1}{2})y^2 - 2\lambda(x - yt) + \lambda^2(1 - t^2)$$

a perfect square. This will obviously be

$$\left(\frac{x - yt}{\sqrt{1 - t^2}} - \lambda\sqrt{1 - t^2}\right)^2,$$

so that
$$A = \frac{1}{2} - \frac{1}{1 - t^2} = -\frac{1 + t^2}{2(1 - t^2)};$$

$$B = \frac{2t}{1 - t^2};$$

$$C = -\frac{1}{2} - \frac{t^2}{1 - t^2} = -\frac{1 + t^2}{2(1 - t^2)},$$

and $\exp(Ax^2 + Bxy + Cy^2) = \exp\left[\dfrac{4xyt - (x^2 + y^2)(1 + t^2)}{2(1 - t^2)}\right].$

As to D, we have

$$1 = \int_{-\infty}^{\infty} D \exp\left\{-\left(\frac{x - yt}{\sqrt{1 - t^2}} - \lambda\sqrt{1 - t^2}\right)^2\right\} dx$$

$$= D \int_{-\infty}^{\infty} e^{-\frac{x^2}{1 - t^2}} dx = D\sqrt{\pi(1 - t^2)},$$

so that
$$D = \frac{1}{\sqrt{\pi(1 - t^2)}}$$

and $K(x, y, t) = \dfrac{1}{\sqrt{\pi(1 - t^2)}} \exp\left[\dfrac{4xyt - (x^2 + y^2)(1 + t^2)}{2(1 - t^2)}\right].$

It remains to verify that this is actually the function $K(x, y, t)$ of (7·06). To begin with, it is easy to verify that

$$\frac{\partial K}{\partial x} = \frac{2yt - x(1 + t^2)}{1 - t^2} K(x, y, t);$$

$$\frac{\partial^2 K}{\partial x^2} = \left\{\left[\frac{2yt - x(1 + t^2)}{1 - t^2}\right]^2 - \frac{1 + t^2}{1 - t^2}\right\} K(x, y, t);$$

PLANCHEREL'S THEOREM

$$\frac{\partial^2 K}{\partial x^2} - x^2 K = \left\{\frac{4y^2 t^2 - 4xyt(1+t^2) + x^2(1+t^2)^2 - (1-t^4) - x^2(1-t^2)^2}{(1-t^2)^2}\right\}$$
$$\times K(x, y, t)$$
$$= \left\{\frac{4(x^2+y^2)t^2 - 4xyt(1+t^2) - (1-t^4)}{(1-t^2)^2}\right\} K(x, y, t).$$

As this expression is symmetrical in x and y, we see that
$$\frac{\partial^2 K}{\partial x^2} - x^2 K = \frac{\partial^2 K}{\partial y^2} - y^2 K.$$
Again,
$$-2t\frac{\partial K}{\partial t} - K = K\left\{-2t\left[\frac{t}{1-t^2} + \frac{2xy(1+t^2)}{(1-t^2)^2} - \frac{2t(x^2+y^2)}{(1-t^2)^2}\right] - 1\right\}$$
$$= \left\{\frac{4(x^2+y^2)t^2 - 4xyt(1+t^2) - (1-t^4)}{(1-t^2)^2}\right\} K(x, y, t)$$
$$= \frac{\partial^2 K}{\partial x^2} - x^2 K = \frac{\partial^2 K}{\partial y^2} - y^2 K.$$

The function $K(x, y, t)$ is analytic in t for $|t| < 1$, for all values of x and y. It is easy to show by mathematical induction that
$$\frac{\partial^n K}{\partial t^n} = P(x, y, t) K(x, y, t),$$
where $P(x, y, t)$ is a polynomial in x and y with coefficients dependent on t. Thus, by Taylor's theorem,

(7·08) $$K(x, y, t) = \sum_0^\infty P_n(x, y) t^n K(x, y, 0)$$
$$= \sum_0^\infty P_n(x, y) t^n e^{-\frac{x^2+y^2}{2}},$$

where the $P_n(x, y)$ are polynomials, not necessarily of the nth degree, in x and y. Hence

(7·09) $$-2t\frac{\partial K}{\partial t} - K = \sum_0^\infty -(2n+1) P_n(x, y) t^n e^{-\frac{x^2+y^2}{2}}.$$

When expanded as a series in powers of t, $K(x, y, t)$ is dominated by the expansion of

(7·10) $$\frac{1}{\sqrt{\pi(1-t^2)}} \exp\left[\frac{4ABt + (A^2+B^2)(1+t^2)}{2(1-t^2)}\right]$$

over $|x| < A, |y| < B$. It is easy to verify that for $|t| < 1$, (7·10) may be expanded in a convergent Taylor series in x and y.

Thus if $|t|<1$, the series (7·08) converges uniformly over any finite complex domain of x and y. It may thus be differentiated term by term any number of times* with respect to x or y, and

$$\frac{\partial^2 K}{\partial x^2} - x^2 K = \sum_0^\infty t^n \left(\frac{\partial^2}{\partial x^2} - x^2\right)\left(P_n(x, y) e^{-\frac{x^2+y^2}{2}}\right);$$

$$\frac{\partial^2 K}{\partial y^2} - y^2 K = \sum_0^\infty t^n \left(\frac{\partial^2}{\partial y^2} - y^2\right)\left(P_n(x, y) e^{-\frac{x^2+y^2}{2}}\right).$$

Thus, by (7·09),

$$\left(\frac{\partial^2}{\partial x^2} - x^2\right)\left(P_n(x,y) e^{-\frac{x^2+y^2}{2}}\right) = \left(\frac{\partial^2}{\partial y^2} - y^2\right)\left(P_n(x,y) e^{-\frac{x^2+y^2}{2}}\right)$$

$$= -(2n+1) P_n(x, y) e^{-\frac{x^2+y^2}{2}}.$$

In view of the equivalence of (6·01) and (6·02), this yields

$$\frac{\partial^2 P_n}{\partial x^2} - 2x \frac{\partial P_n}{\partial x} + 2n P_n = \frac{\partial^2 P_n}{\partial y^2} - 2y \frac{\partial P_n}{\partial y} + 2n P_n = 0.$$

However, since $P_n(x, y)$ is a polynomial in x with coefficients depending on y, and also a polynomial in y with coefficients depending on x, it follows from our discussion of the polynomial solutions of (6·04), that

$$P_n(x, y) = H_n(x) F(y) = H_n(y) G(x),$$

where $F(y)$ and $G(x)$ are functions yet to be determined. Thus

$$G(x)/H_n(x) = F(y)/H_n(y),$$

and the common value of these two expressions can contain neither x nor y, and must be a constant. Hence

$$P_n(x, y) = c_n H_n(x) H_n(y),$$

and

(7·11) $\qquad K(x, y, t) = \sum_0^\infty d_n t^n \psi_n(x) \psi_n(y).$

Here c_n and d_n are functions of n only.

In particular,

(7·12) $\quad K(x, x, t) = \dfrac{1}{\sqrt{\pi(1-t^2)}} e^{-x^2 \frac{1-t}{1+t}} = \sum_0^\infty d_n t^n [\psi_n(x)]^2,$

* Cf. Osgood, *Lehrbuch der Funktionentheorie*, II, i, p. 13, Satz 1.

PLANCHEREL'S THEOREM

and even more specially,

(7·13) $\quad K(0, 0, t) = \pi^{-\frac{1}{2}}\left(1 + \dfrac{t^2}{2} + \dfrac{1.3}{2^3} t^4 + \dfrac{1.3.5}{2^5} t^6 + \ldots\right)$

$\qquad = \sum\limits_{0}^{\infty} d_n t^n [\psi_n(0)]^2.$

We have already seen that $\psi_n(0)$ does not vanish for any even n, and the first series of (7·13) has positive even terms. Hence no even d_n is negative.

Again, we have already seen that (7·11) may be differentiated term by term, so that

$\dfrac{\partial^2}{\partial x \partial y} K(x, y, t) = \sum\limits_{0}^{\infty} d_n t^n \psi_n'(x) \psi_n'(y)$

$\qquad = \left\{\left(\dfrac{2yt - x(1+t^2)}{1-t^2}\right)\left(\dfrac{2xt - y(1+t^2)}{1-t^2}\right) + \dfrac{2t}{1-t^2}\right\}$

$\qquad\qquad\qquad\qquad\qquad\qquad \times K(x, y, t).$

Thus $\quad \sum\limits_{0}^{\infty} d_n t^n [\psi_n'(0)]^2 = K(0, 0, t) [2t + 2t^3 + 2t^5 + \ldots],$

and being the product of two series without negative terms, contains no powers of t with a negative coefficient. Also, $\psi_n'(0)$ vanishes for no odd n, so that no odd d_n is negative. Thus no d_n is negative, and the series of (7·12) contains only non-negative terms for $0 \leqslant t < 1$. Thus (7·12) may be integrated term by term, by monotone convergence, and as the set $\{\psi_n\}$ is normal, we obtain

$\sum\limits_{0}^{\infty} d_n t^n = \sum\limits_{0}^{\infty} d_n t^n \int_{-\infty}^{\infty} [\psi_n(x)]^2 dx$

$\qquad = \dfrac{1}{\sqrt{\pi(1-t^2)}} \int_{-\infty}^{\infty} e^{-x^2 \frac{1-t}{1+t}} dx$

$\qquad = \dfrac{1}{1-t} \sqrt{\dfrac{1-t}{\pi(1+t)}} \int_{-\infty}^{\infty} e^{-x^2 \frac{1-t}{1+t}} dx$

$\qquad = \dfrac{1}{1-t} \dfrac{1}{\sqrt{\pi}} \int_{-\infty}^{\infty} e^{-y^2} dy$

$\qquad = \dfrac{1}{1-t}$

$\qquad = \sum\limits_{0}^{\infty} t^n.$

62 PLANCHEREL'S THEOREM

That is, every d_n is 1, and we have proved that

$$(7\cdot 14) \quad K(x, y, t) = \frac{1}{\sqrt{\pi(1-t^2)}} \exp\left[\frac{4xyt - (x^2+y^2)(1+t^2)}{2(1-t^2)}\right]$$

$$= \sum_{0}^{\infty} t^n \psi_n(x) \psi_n(y).$$

Now, we have seen that the theory of normal and orthogonal functions of two variables differs in no way from that of normal and orthogonal functions of one variable. The functions $\psi_m(x) \psi_n(y)$ with the double index (m, n) are a normal and orthogonal set over the entire plane, for

$$\int_{-\infty}^{\infty} dx \int_{-\infty}^{\infty} \psi_m(x) \psi_n(y) \psi_\mu(x) \psi_\nu(y) \, dy$$

$$= \int_{-\infty}^{\infty} \psi_m(x) \psi_\mu(x) \, dx \int_{-\infty}^{\infty} \psi_n(y) \psi_\nu(y) \, dy$$

$$\begin{cases} = 1 & \text{if } m = \mu, \, n = \nu; \\ = 0 & \text{otherwise.} \end{cases}$$

Let $|t| < 1$. Since the series $\sum_{0}^{\infty} |t^{2n}|$ converges, it follows by the Riesz-Fischer theorem that the series of (7·14) converges in the mean in (x, y) and hence, by X$_{41}$, that it converges in the mean to $K(x, y, t)$. That is,

$$\lim_{n \to \infty} \int_{-\infty}^{\infty} dx \int_{-\infty}^{\infty} dy \left| K(x, y, t) - \sum_{0}^{n} t^k \psi_k(x) \psi_k(y) \right|^2 = 0.$$

Now let $\sum_{0}^{\infty} \epsilon_\nu$ be a convergent sequence, and let n_ν be so large that

$$\int_{-\infty}^{\infty} dx \int_{-\infty}^{\infty} dy \left| K(x, y, t) - \sum_{0}^{n_\nu} t^k \psi_k(x) \psi_k(y) \right|^2 < \epsilon_\nu^2.$$

It follows from X$_{25}$ that there exists a set S_ν of values of x, of measure not exceeding ϵ_ν, such that, if x is outside S_ν,

$$\int_{-\infty}^{\infty} \left| K(x, y, t) - \sum_{0}^{n_\nu} t^k \psi_k(x) \psi_k(y) \right|^2 dy < \epsilon_\nu.$$

This means again that outside the set $S_\nu + S_{\nu+1} + \ldots = T_\nu$, of measure not exceeding $\sum_{\nu}^{\infty} \epsilon_\mu$,

$$(7\cdot 15) \quad \lim_{\mu \to \infty} \int_{-\infty}^{\infty} \left| K(x, y, t) - \sum_{0}^{n_\mu} t^k \psi_k(x) \psi_k(y) \right|^2 dy = 0,$$

PLANCHEREL'S THEOREM

and since
$$\lim_{\nu \to \infty} \sum_{\nu}^{\infty} \epsilon_\mu = 0,$$

that (7·15) is valid except on the null set common to all T_ν's. If we consider $K(x, y, t)$ as a function of y alone, and apply the Bessel inequality to (7·15), we obtain

$$\sum_0^\infty t^n [\psi_n(x)]^2 < \infty$$

for almost all x. Thus the series of (7·14) converges in the mean in y, for almost all x. By the principle X_{41} that if a series converges to one function and converges in the mean to another, these can differ at the most over a null set, it follows that for almost all x, and considered as a function of y,

$$K(x, y, t) = \underset{n \to \infty}{\text{l.i.m.}} \sum_0^n t^k \psi_k(x) \psi_k(y).$$

Thus by X_{47} and the Parseval theorem X_{52}, if $f(x)$ belongs to L_2, we have for almost all x,

$$(7 \cdot 16) \quad \int_{-\infty}^{\infty} f(y) K(x, y, t) \, dy$$
$$= \sum_0^\infty t^n \psi_n(x) \int_{-\infty}^{\infty} f(y) \psi_n(y) \, dy.$$

By the Bessel inequality,

$$\sum_0^\infty \left| \int_{-\infty}^{\infty} f(y) \psi_n(y) \, dy \right|^2 \leqslant \int_{-\infty}^{\infty} |f(x)|^2 \, dx < \infty.$$

Therefore
$$\sum_0^\infty \left| t^n \int_{-\infty}^{\infty} f(y) \psi_n(y) \, dy \right|^2 < \infty,$$

and by the Riesz-Fischer theorem X_{46}, there exists a function $F(x, t)$ belonging to L_2, and such that

$$(7 \cdot 17) \quad F(x, t) \sim \sum_0^\infty t^n \psi_n(x) \int_{-\infty}^{\infty} f(y) \psi_n(y) \, dy.$$

By another application of X_{41} to (7·16) and (7·17), we obtain

$$(7 \cdot 18) \quad \int_{-\infty}^{\infty} f(y) K(x, y, t) \, dy \sim \sum_0^\infty t^n \psi_n(x) \int_{-\infty}^{\infty} f(y) \psi_n(y) \, dy.$$

§ 8. The Closure of the Hermite Functions.

On the basis of (7·18), we wish to prove that if $f(x)$ belongs to L_2,

$$f(x) \sim \sum_0^\infty \psi_n(x) \int_{-\infty}^\infty f(y)\psi_n(y)\,dy,$$

or in other words,

Theorem 1. *The Hermite functions are a complete normal set.*

By X_{50} we shall have established this if we can show that, given $\epsilon > 0$, if f is a step-function belonging to L_2, we can find an N and a t such that

$$\int_{-\infty}^\infty \left| f(x) - \sum_0^N t^n \psi_n(x) \int_{-\infty}^\infty f(y)\psi_n(y)\,dy \right|^2 dx < \epsilon.$$

Proposition (7·18) and the Minkowski inequality show that this will be possible if

$$(8\cdot01) \quad \lim_{t \to 1-0} \int_{-\infty}^\infty \left| f(x) - \int_{-\infty}^\infty f(y) K(x,y,t)\,dy \right|^2 dy = 0.$$

We shall now establish (8·01) for any step-function $f(x)$ belonging to L_2, and shall thus complete the proof of theorem 1.

Over any interval (a, b) containing x as an interior point,

$$(8\cdot012) \quad \lim_{t \to 1-0} \int_a^b K(x,y,t)\,dy = \lim_{t \to 1-0} \frac{1}{\sqrt{\pi(1-t^2)}}$$

$$\times \int_a^b \exp\left[\frac{-\left(y\sqrt{1+t^2} - 2x\dfrac{t}{\sqrt{1+t^2}}\right)^2}{2(1-t^2)} - \frac{a^2(1-t^2)}{2(1+t^2)} \right] dy$$

$$= \lim_{t \to 1-0} \frac{1}{\sqrt{2\pi(1-t)}} \int_{a-x}^{b-x} e^{-\frac{y^2}{2(1-t)}}\,dy = 1.$$

$K(x, y, t)$ is clearly non-negative. Again, let us put

$$x + y = \xi, \quad x - y = \eta.$$

Then

$$(8\cdot013)$$
$$K(x,y,t) = \frac{1}{\sqrt{\pi(1-t^2)}} \exp\left[\frac{(\xi^2 - \eta^2)t - \dfrac{\xi^2 + \eta^2}{2}(1+t^2)}{2(1-t^2)} \right]$$

$$= \frac{1}{\sqrt{\pi(1-t^2)}} \exp\left[-\frac{\xi^2(1-t)}{4(1+t)} - \frac{\eta^2(1+t)}{4(1-t)} \right].$$

Hence
$$\lim_{t \to 1-0} K(x, y, t) = 0$$
uniformly for all x and y in (a, b) for which $|y - x| > \epsilon$. We may thus apply X_{55}, which yields us

$$f(x) = \lim_{t \to 1-0} \int_a^b f(y) K(x, y, t) \, dy$$

boundedly for the points of the interior of the interval (a, b) at which $f(x)$ is continuous. However, $f(x)$ is a step-function belonging to L_2, it vanishes for large enough arguments, and has only a finite set of points of discontinuity. Thus

$$f(x) = \lim_{t \to 1-0} \int_{-\infty}^{\infty} f(y) K(x, y, t) \, dy$$

at every point of continuity, as we may see by taking $-a$ and b so large that $f(x)$ vanishes outside (a, b). The expression

$$\int_{-\infty}^{\infty} f(y) K(x, y, t) \, dy$$

is bounded in (x, t) over any finite interval of x and for $0 \leqslant t \leqslant 1$. Thus, over any *finite* interval of x,

(8·02) $\qquad f(x) = \underset{t \to 1-0}{\text{l.i.m.}} \int_{-\infty}^{\infty} f(y) K(x, y, t) \, dy.$

We now shall show that if $f(x)$ differs from 0 only for $|x| < A$, then (8·02) will hold over $(-\infty, -2A)$ and over $(2A, \infty)$. Over these ranges, (8·02) becomes

(8·025) $\qquad \underset{t \to 1-0}{\text{l.i.m.}} \int_{-\infty}^{\infty} f(y) K(x, y, t) \, dy = 0,$

which is what we must prove. However, over the ranges in question, by (8·013),

(8·03) $\left| \int_{-\infty}^{\infty} f(y) K(x, y, t) \, dy \right|$
$\leqslant \int_{-\infty}^{\infty} |f(y)| \, dy \, \frac{1}{\sqrt{\pi(1-t^2)}} \exp\left[-\frac{(x/2)^2(1-t)}{4(1+t)} - \frac{A^2(1+t)}{4(1-t)} \right],$

since, if $f(y) \neq 0$,
$$|\xi| = |x + y| \geqslant |x| - A \geqslant |x| - |x|/2 = |x|/2;$$
$$|\eta| = |x - y| \geqslant A.$$

PLANCHEREL'S THEOREM

The right-hand expression in (8·03) is of the form
$$\psi(t)\,\phi(t)\,e^{-x^2[\phi(t)]^2},$$
where $[\psi(t)]^2\,\phi(t)$ tends to 0 as $t\to 1-0$.

Thus
$$\left[\int_{-\infty}^{-2A}+\int_{2A}^{\infty}\right]\left|\int_{-\infty}^{\infty}f(y)\,K(x,y,t)\,dy\right|^2 dx$$
$$\leqslant [\psi(t)]^2 \int_{-\infty}^{\infty}[\phi(t)]^2\,e^{-x^2[\phi(t)]^2}\,dx$$
$$\leqslant \text{const.}\,[\psi(t)]^2\,\phi(t),$$
from which (8·025) follows at once. Thus (8·02) holds over $(-\infty,-2A)$, $(-2A,2A)$, and $(2A,\infty)$, or what is the same thing, over $(-\infty,\infty)$. This is however equivalent to (8·01), which we have shown to be equivalent to theorem 1, which we have thus proved.

By (7·14), if $f(x)$ belongs to L_2,

(8·04) $\quad \dfrac{1}{\sqrt{\pi(1+t^2)}}\displaystyle\int_{-\infty}^{\infty}f(y)\exp\left[\dfrac{-4i\,xyt-(x^2+y^2)(1-t^2)}{2(1+t^2)}\right]dy$

$$=\int_{-\infty}^{\infty}f(y)\,K(x,y,-it)\,dy$$

$$=\underset{N\to\infty}{\text{l.i.m.}}\sum_{0}^{N}(-it)^n\,\psi_n(x)\int_{-\infty}^{\infty}f(y)\,\psi_n(y)\,dy.$$

By the Bessel inequality,
$$\infty > \sum_{0}^{\infty}\left|\int_{-\infty}^{\infty}f(y)\,\psi_n(y)\,dy\right|^2 = \sum_{0}^{\infty}\left|(-i)^n\int_{-\infty}^{\infty}f(y)\,\psi_n(y)\,dy\right|^2.$$

Thus, by the Riesz-Fischer theorem, there exists a function $g(x)$ belonging to L_2, such that
$$g(x)\sim\sum_{0}^{\infty}(-i)^n\psi_n(x)\int_{-\infty}^{\infty}f(y)\,\psi_n(y)\,dy.$$

Since the set $\{\psi_n\}$ is complete or closed, we have
$$f(x)\sim\sum_{0}^{\infty}\psi_n(x)\int_{-\infty}^{\infty}f(y)\,\psi_n(y)\,dy.$$

Thus two applications of the Hurwitz-Parseval theorem give us

(8·05)
$$\int_{-\infty}^{\infty}|g(x)|^2\,dx = \sum_{0}^{\infty}\left|\int_{-\infty}^{\infty}f(y)\,\psi_n(y)\,dy\right|^2 = \int_{-\infty}^{\infty}|f(x)|^2\,dx.$$

PLANCHEREL'S THEOREM

This theorem also gives us, for $0 \leqslant t < 1$,

$$\int_{-\infty}^{\infty} \left| g(x) - \frac{1}{\sqrt{\pi(1+t^2)}} \int_{-\infty}^{\infty} f(y) \right.$$
$$\left. \exp\left[\frac{-4ixyt - (x^2+y^2)(1-t^2)}{2(1+t^2)}\right] dy \right|^2 dx$$
$$= \sum_{0}^{\infty} (1-t^n) \left| \int_{-\infty}^{\infty} f(y) \psi_n(y) \, dy \right|^2.$$

From this and (8·05), it results that

$$\overline{\lim_{t \to 1-0}} \int_{-\infty}^{\infty} \left| g(x) - \frac{1}{\sqrt{\pi(1+t^2)}} \int_{-\infty}^{\infty} f(x) \right.$$
$$\left. \exp\left[\frac{-4ixyt - (x^2+y^2)(1-t^2)}{2(1+t^2)}\right] dy \right|^2 dx$$
$$\leqslant \overline{\lim} \sum_{0}^{N} (1-t^n) \left| \int f \psi_n \right|^2 + \sum_{N}^{\infty} \left| \int f \psi_n \right|^2 \leqslant \sum_{N}^{\infty} \left| \int f \psi_n \right|^2.$$

Since the series in (8·05) converges, it further results that

(8·06) $\quad g(x) = \underset{t \to 1-0}{\text{l.i.m.}} \frac{1}{\sqrt{\pi(1+t^2)}} \int_{-\infty}^{\infty} f(y)$
$$\exp\left[\frac{-4ixyt - (x^2+y^2)(1-t^2)}{2(1+t^2)}\right] dy.$$

§ 9. The Fourier Transform.

We now wish to simplify the definition of the function given by (8·06). In case $f(x) = {}_A f(x)$, and vanishes for large arguments, we have a finite range of integration for the integral in (8·06). As $t \to 1$, the integrand is dominated by an expression of the form $c|f(y)|$, and tends almost everywhere to $f(y) e^{-ixy}$. Thus, for such a function,

(9·01) $\qquad g(x) = \frac{1}{\sqrt{2\pi}} \int_{-\infty}^{\infty} f(y) e^{-ixy} dy.$

Now let $f(x)$ be *any* function of L_2, and let us submit $f(x) - {}_A f(x)$ to the linear transformation given by (8·06). By (9·01) this will yield us

$$g(x) - \frac{1}{\sqrt{2\pi}} \int_{-A}^{A} f(y) e^{-ixy} dy,$$

and by (8·05) we shall have

$$\int_{-\infty}^{\infty} \left| g(x) - \frac{1}{\sqrt{2\pi}} \int_{-A}^{A} f(y) e^{-ixy} dy \right|^2 dx = \left[\int_{-\infty}^{-A} + \int_{A}^{\infty} \right] |f(x)|^2 dx.$$

It then follows from the convergence of $\int_{-\infty}^{\infty} |f(x)|^2 dx$ that

(9·02) $\quad \lim_{A \to \infty} \int_{-\infty}^{\infty} \left| g(x) - \frac{1}{\sqrt{2\pi}} \int_{-A}^{A} f(y) e^{-ixy} dy \right|^2 dx = 0,$

or

(9·03) $\quad g(x) = \underset{A \to \infty}{\text{l.i.m.}} \frac{1}{\sqrt{2\pi}} \int_{-A}^{A} f(y) e^{-ixy} dy$

$$\sim \sum_{0}^{\infty} (-i)^n \psi_n(x) \int_{-\infty}^{\infty} f(y) \psi_n(y) dy.$$

Clearly $\int_{-\infty}^{\infty} g(x) \psi_n(x) dx = (-i)^n \int_{-\infty}^{\infty} f(y) \psi_n(y) dy,$

and $\quad f(x) \sim \sum_{0}^{\infty} i^n \psi_n(x) \int_{-\infty}^{\infty} g(y) \psi_n(y) dy.$

As in (8·04), if we replace $f(x)$ by $g(x)$ and i by $-i$, we have

$$\sum_{0}^{\infty} (it)^n \psi_n(x) \int_{-\infty}^{\infty} g(y) \psi_n(y) dy$$

$$= \frac{1}{\sqrt{\pi(1+t^2)}} \int_{-\infty}^{\infty} g(y) \exp \left[\frac{4ixyt - (x^2 + y^2)(1 - t^2)}{2(1 + t^2)} \right] dy$$

and an argument in all respects like that by which we deduced (8·06) will show that

(9·04) $\quad f(x) = \underset{t \to 1-0}{\text{l.i.m.}} \frac{1}{\sqrt{\pi(1+t^2)}} \int_{-\infty}^{\infty} g(y)$

$$\exp \left[\frac{4ixyt - (x^2 + y^2)(1 - t^2)}{2(1 + t^2)} \right] dy.$$

Just as (9·03) follows from (8·06) so from (9·04) we may deduce

(9·05) $\quad f(x) = \underset{A \to \infty}{\text{l.i.m.}} \frac{1}{\sqrt{2\pi}} \int_{-A}^{A} g(y) e^{ixy} dy.$

Combining (9·03), (8·05), and (9·05), we obtain

PLANCHEREL'S THEOREM

Theorem 2. (*Plancherel's Theorem.*) *If $f(x)$ belongs to L_2 on $(-\infty, \infty)$, then the function*

$$(9\cdot06) \qquad g(x) = \underset{A\to\infty}{\text{l.i.m.}} \frac{1}{\sqrt{2\pi}} \int_{-A}^{A} f(y) e^{-ixy} \, dy,$$

known as the Fourier transform of $f(x)$, exists and belongs to L_2. We have

$$(8\cdot05) \qquad \int_{-\infty}^{\infty} |g(x)|^2 \, dx = \int_{-\infty}^{\infty} |f(x)|^2 \, dx,$$

and
$$f(x) = \underset{A\to\infty}{\text{l.i.m.}} \frac{1}{\sqrt{2\pi}} \int_{-A}^{A} g(y) e^{ixy} \, dy.$$

By the Schwarz inequality, it results from (9·02) that

$$\lim_{A\to\infty} \left| \int_0^x g(x) \, dx - \frac{1}{\sqrt{2\pi}} \int_0^x dx \int_{-A}^A f(y) e^{-ixy} \, dy \right| = 0.$$

Hence, by X_{24},

$$\int_0^x g(x) \, dx = \frac{1}{\sqrt{2\pi}} \int_{-\infty}^{\infty} f(y) \, dy \int_0^x e^{-ixy} \, dx$$

$$= \frac{1}{\sqrt{2\pi}} \int_{-\infty}^{\infty} f(y) \frac{1 - e^{-ixy}}{iy} \, dy,$$

and, by X_{13},

$$(9\cdot07) \qquad g(x) = \frac{1}{\sqrt{2\pi}} \frac{d}{dx} \int_{-\infty}^{\infty} f(y) \frac{1 - e^{-ixy}}{iy} \, dy$$

almost everywhere. Similarly,

$$(9\cdot08) \qquad f(x) = \frac{1}{\sqrt{2\pi}} \frac{d}{dx} \int_{-\infty}^{\infty} g(y) \frac{e^{ixy} - 1}{iy} \, dy$$

almost everywhere. If $f(x)$ is even, (9·07) and (9·08) reduce to

$$g(x) = \sqrt{\frac{2}{\pi}} \frac{d}{dx} \int_0^{\infty} f(y) \frac{\sin xy}{y} \, dy,$$

and
$$f(x) = \sqrt{\frac{2}{\pi}} \frac{d}{dx} \int_0^{\infty} g(y) \frac{\sin xy}{y} \, dy,$$

and, if $f(x)$ is odd, these are replaced by

$$ig(x) = \sqrt{\frac{2}{\pi}} \frac{d}{dx} \int_0^{\infty} f(y) \frac{1 - \cos xy}{y} \, dy,$$

70   PLANCHEREL'S THEOREM

and $\quad f(x) = \sqrt{\dfrac{2}{\pi}} \dfrac{d}{dx} \displaystyle\int_0^\infty ig(y) \dfrac{1-\cos xy}{y} dy.$

These are the formulae originally given by Plancherel. In the even case, $g(x)$ and $f(x)$ are said to be cosine transforms of each other, and in the odd case $ig(x)$ and $f(x)$ are said to be sine transforms of each other.

The method of proof of the Plancherel theorem given in this chapter differs considerably from that originally given by Plancherel*. Other proofs have been given by Titchmarsh†, F. Riesz‡, and the author§. The present method is based on a suggestion of Campbell‖, and involves an argument concerning the Abel summability of a development in Hermite polynomials closely allied to one given by Müntz¶. A somewhat similar discussion of Hermite developments is to be found in an earlier paper by the author**.

By methods precisely similar to those used in the deduction of X_{52} from X_{51}, we obtain by theorem 2:

THEOREM 3. (*The Parseval theorem for the Fourier integral.*) *If $g_1(x)$ is the Fourier transform of $f_1(x)$ and $g_2(x)$ is the Fourier transform of $f_2(x)$, then*

$$\int_{-\infty}^{\infty} g_1(x)\overline{g_2(x)}\, dx = \int_{-\infty}^{\infty} f_1(x)\overline{f_2(x)}\, dx.$$

Hence, as in X_{61},

$$\int_{-\infty}^{\infty} g_1(x) g_2(x)\, dx = \int_{-\infty}^{\infty} f_1(x) f_2(-x)\, dx;$$

(9·09) $\quad \displaystyle\int_{-\infty}^{\infty} g_1(x) g_2(y-x)\, dx = \int_{-\infty}^{\infty} f_1(x) f_2(x) e^{-iyx}\, dx,$

$$\int_{-\infty}^{\infty} g_1(x) g_2(x) e^{iyx}\, dy = \int_{-\infty}^{\infty} f_1(x) f_2(y-x)\, dx.$$

* Plancherel 1.   † Titchmarsh 1.   ‡ F. Riesz 2.
§ Wiener 1.   ‖ Campbell and Foster 1.
¶ Ch. H. Müntz, "Über die Potenzsummation einer Entwickelung nach Hermiteschen Polynomen," *Math. Ztschr.* 31 (1930), 350. The K of (7·06) had already made its appearance in a paper of Mehler (*Journ. für Math.* 66 (1866), 174.   ** Wiener 3.

PLANCHEREL'S THEOREM

By (9·09) the Fourier transform of the product of two functions f_1 and f_2 is $\sqrt{2\pi}$ times the "Faltung"

$$\int_{-\infty}^{\infty} g_1(x) g_2(y-x)$$

of their Fourier transforms. It results from X_{41} that if furthermore $f_1(x) f_2(x)$ belongs to L_2, this statement is true in fact as well as formally, for by X_{41} the equation

$$\int_{-\infty}^{\infty} f_1(x) f_2(x) e^{-iyx} dx = \underset{A \to \infty}{\text{l.i.m.}} \int_{-A}^{A} f_1(x) f_2(x) e^{-iyx} dx$$

is true almost everywhere if the left-hand side exists for almost all y and the right-hand side exists, as will be the case, since $f_1(x) f_2(x)$ belongs to L_2.

CHAPTER II

THE GENERAL TAUBERIAN THEOREM

§10. **Enunciation of the General Tauberian Theorem.**

In Chapter I we discussed functions belonging to L_2 over $(-\infty, \infty)$. In the present chapter we are chiefly concerned with functions $f(x)$ belonging to L_1 over this interval. If $f(x)$ is such a function,

$$g(u) = \frac{1}{\sqrt{2\pi}} \int_{-\infty}^{\infty} f(x) e^{-iux} dx$$

exists for every u. It is bounded, and indeed

$$|g(u)| \leq \frac{1}{\sqrt{2\pi}} \int_{-\infty}^{\infty} |f(x)| dx.$$

It is as a matter of fact uniformly continuous, for

$$|g(u+\epsilon) - g(u)| = \frac{1}{\sqrt{2\pi}} \left| \int_{-\infty}^{\infty} f(x) e^{-iux} (e^{-ix\epsilon} - 1) dx \right|$$

$$\leq \frac{1}{\sqrt{2\pi}} \int_{-\infty}^{\infty} |f(x)| \, 2 \left|\sin \frac{x\epsilon}{2}\right| dx$$

$$\leq \frac{1}{\sqrt{2\pi}} \left[\int_{-\infty}^{-A} + \int_{A}^{\infty} \right] 2 |f(x)| dx + \frac{\epsilon A}{\sqrt{2\pi}} \int_{-A}^{A} |f(x)| dx.$$

Now let A be so large that

$$\frac{1}{\sqrt{2\pi}} \left[\int_{-\infty}^{-A} + \int_{A}^{\infty} \right] 2 |f(x)| dx < \eta/2.$$

Let us then choose ϵ so small that

$$\frac{\epsilon A}{\sqrt{2\pi}} \int_{-A}^{A} |f(x)| dx < \eta/2.$$

We shall then have

$$|g(u+\epsilon) - g(u)| < \eta.$$

We shall term $g(u)$ the *Fourier transform* of $f(x)$. The class L_1 differs from L_2 in that the Fourier transform of a function of

THE GENERAL TAUBERIAN THEOREM

L_1 is defined for *every* real argument, and not merely for almost every real argument.

We shall also introduce a sub-class M_1 of L_1, consisting of all continuous functions $f(x)$ for which for some positive a and real b,

$$(10\cdot01) \qquad \sum_{n=-\infty}^{\infty} \max_{na+b \leqslant x \leqslant (n+1)a+b} |f(x)| < \infty.$$

It is clear that if (10·01) is true for any positive a and real b, it is true for any other positive a_1 and real b_1. This follows from the fact that if $[a_1/a] = \nu$, no interval of the form

$$(na_1 + b_1,\ (n+1)a_1 + b_1)$$

contains parts of more than $\nu + 2$ intervals of the form

$$(ma + b,\ (m+1)a + b).$$

Thus

$$(10\cdot02) \qquad \sum_{n=-\infty}^{\infty} \max_{na_1+b_1 \leqslant x \leqslant (n+1)a_1+b_1} |f(x)|$$

$$\geqslant \frac{1}{\nu + 2} \sum_{n=-\infty}^{\infty} \max_{na+b \leqslant x \leqslant (n+1)a+b} |f(x)|.$$

Similarly, if $[a/a_1] = \mu$,

$$\sum_{n=-\infty}^{\infty} \max_{na+b \leqslant x \leqslant (n+1)a+b} |f(x)| \geqslant \frac{1}{\mu+2} \sum_{n=-\infty}^{\infty} \max_{na_1+b_1 \leqslant x \leqslant (n+1)a_1+b_1} |f(x)|.$$

Thus there is no change in the definition of L_1 if we take $a = 1$, $b = 0$, and replace (10·01) by

$$\sum_{n=-\infty}^{\infty} \max_{n \leqslant x \leqslant n+1} |f(x)| < \infty.$$

Every function of M_1 is of course a member of L_1, and has a bounded uniformly continuous Fourier transform.

We shall devote the present chapter to the demonstration of the two following theorems, or rather variants of one theorem, which we shall term collectively the general Tauberian theorem. These are:

THEOREM 4. *Let $K_1(x)$ belong to L_1, and let its Fourier transform vanish for no real argument. Let $K_2(x)$ belong to L_1. Let $f(x)$ be bounded over $(-\infty, \infty)$. Let*

$$(10\cdot03) \qquad \lim_{x \to \infty} \int_{-\infty}^{\infty} K_1(x-y) f(y)\, dy = A \int_{-\infty}^{\infty} K_1(x)\, dx.$$

74 THE GENERAL TAUBERIAN THEOREM

Then

(10·04) $\quad \lim\limits_{x \to \infty} \int_{-\infty}^{\infty} K_2(x-y) f(y) \, dy = A \int_{-\infty}^{\infty} K_2(x) \, dx.$

Conversely, let $K_1(x)$ belong to L_1, but let its Fourier transform have a real zero. Then there exist a bounded function $f(x)$ and a function $K_2(x)$ belonging to L_1, such that (10·03) is true but (10·04) false.

THEOREM 5. *Let $K_1(x)$ belong to M_1, and let its Fourier transform vanish for no real argument. Let $K_2(x)$ belong to M_1. Let $f(x)$ be of limited total variation over every finite range, and let*

(10·05) $\quad\quad\quad\quad\quad \int_{n}^{n+1} |dg(x)|$

be bounded for $-\infty < n < \infty$. Let

(10·06) $\quad \lim\limits_{x \to \infty} \int_{-\infty}^{\infty} K_1(x-y) \, dg(y) = A \int_{-\infty}^{\infty} K_1(x) \, dx.$

Then

(10·07) $\quad \lim\limits_{x \to \infty} \int_{-\infty}^{\infty} K_2(x-y) \, dg(y) = A \int_{-\infty}^{\infty} K_2(x) \, dx.$

Conversely, let $K_1(x)$ belong to M_1, but let its Fourier transform have a real zero. Then there exists a function $g(x)$ of limited total variation over every finite interval, for which (10·05) is bounded, and a function $K_2(x)$ belonging to M_1, such that (10·06) is true but (10·07) false.

The converse portions of theorems 4 and 5 are quite trivial. Let

$$\frac{1}{\sqrt{2\pi}} \int_{-\infty}^{\infty} K_1(x) e^{-iu_0 x} \, dx = 0.$$

There is no difficulty in finding a function K_2 belonging to M_1 and hence to L_1 for which

$$\frac{1}{\sqrt{2\pi}} \int_{-\infty}^{\infty} K_2(x) e^{-iu_0 x} \, dx = I \neq 0.$$

The function $e^{-x^2/2}$ is an instance in point. If we put

$$f(x) = e^{iu_0 x}, \quad g(x) = e^{iu_0 x} / iu_0$$

THE GENERAL TAUBERIAN THEOREM 75

(10·03) and (10·06) will hold true for $A = 0$, while

$$\int_{-\infty}^{\infty} K_2(x-y) f(y) \, dy = \int_{-\infty}^{\infty} K_2(x-y) \, dg(y) = e^{iu_0 x} I \sqrt{2\pi},$$

which does not tend to any limit whatever. The function $f(x)$ is clearly bounded, while

$$\int_n^{n+1} |dg(x)| = \int_n^{n+1} |e^{iu_0 x}| \, dx = 1.$$

In the proof of the direct parts of theorems 4 and 5, we introduce the notion of the *extension* of a class of functions Σ of L_1 or M_1.

If Σ is a class of functions of L_1, $\epsilon(\Sigma)$, the L_1 extension of Σ, is the class of all functions $K_2(x)$ belonging to L_1, for which, whenever $f(x)$ is bounded and (10·03) holds for every function $K_1(x)$ belonging to Σ, (10·04) holds. Similarly, if Σ is a class of functions of M_1, $\epsilon'(\Sigma)$, the M_1 extension of Σ, is the class of all functions $K_2(x)$ belonging to M_1, for which, whenever $g(x)$ is of limited total variation over every finite range, (10·05) is bounded, and (10·06) holds for every function $K_1(x)$ belonging to Σ, (10·07) holds. The direct parts of theorems 4 and 5 are then special cases of the more general theorems:

THEOREM 6. *If Σ is a class of functions of L_1, and if there is no real argument for which the Fourier transform of every function of Σ vanishes, then*

$$\epsilon(\Sigma) \equiv L_1.$$

THEOREM 7. *If Σ is a class of functions of M_1, and if there is no real argument for which the Fourier transform of every function of Σ vanishes, then*

$$\epsilon'(\Sigma) \equiv M_1.$$

For these theorems we need a number of lemmas. The following lemmas are in part obvious, and no proof of these is given. It is always understood that in a lemma involving ϵ, the functions considered belong to L_1, and in a lemma concerning ϵ', they belong to M_1, unless the opposite is stated.

Lemma 6_1. $\epsilon(\epsilon(\Sigma)) \equiv \epsilon(\Sigma)$.

Lemma 7_1. $\epsilon'(\epsilon'(\Sigma)) \equiv \epsilon'(\Sigma)$.

Lemma 6_2. *If $K(x)$ and $Q(x)$ belong to Σ, $K(x) + Q(x)$ belongs to $\epsilon(\Sigma)$.*

Lemma 7_2. *If $K(x)$ and $Q(x)$ belong to Σ, $K(x) + Q(x)$ belongs to $\epsilon'(\Sigma)$.*

Lemma 6_3. *If A is a constant and $K(x)$ belongs to Σ, $AK(x)$ belongs to $\epsilon(\Sigma)$.*

Lemma 7_3. *If A is a constant and $K(x)$ belongs to Σ, $AK(x)$ belongs to $\epsilon'(\Sigma)$.*

Lemma 6_4. *If a is a real constant and $K(x)$ belongs to Σ, $K(x+a)$ belongs to $\epsilon(\Sigma)$.*

Lemma 7_4. *If a is a real constant and $K(x)$ belongs to Σ, $K(x+a)$ belongs to $\epsilon'(\Sigma)$.*

Lemma 6_5. *If $K(x)$ belongs to Σ and $Q(x)$ belongs to L_1,*

$$(10\cdot08) \qquad \int_{-\infty}^{\infty} K(x-\xi)Q(\xi)\,d\xi$$

belongs to $\epsilon(\Sigma)$.

To prove this lemma, let us first note that

$$\int_{-\infty}^{\infty} K(x-\xi)Q(\xi)\,d\xi$$

belongs to L_1, and in fact that

$$(10\cdot09) \quad \int_{-\infty}^{\infty}\left|\int_{-\infty}^{\infty} K(x-\xi)Q(\xi)\,d\xi\right|dx \leq \int_{-\infty}^{\infty}|K(x)|\,dx\int_{-\infty}^{\infty}|Q(\xi)|\,d\xi.$$

Again, (10·03) yields

$$(10\cdot10) \quad \int_{-\infty}^{\infty} Q(\xi)\,d\xi \lim_{x\to\infty}\int_{-\infty}^{\infty} K_1(x-\xi-y)f(y)\,dy$$

$$= A\int_{-\infty}^{\infty} Q(\xi)\,d\xi\int_{-\infty}^{\infty} K_1(x)\,dx$$

$$= A\int_{-\infty}^{\infty} dx\int_{-\infty}^{\infty} K(x-\xi)Q(\xi)\,d\xi,$$

THE GENERAL TAUBERIAN THEOREM 77

as the integral converges absolutely. Furthermore, since f is bounded and K belongs to L_1,

(10·11) $$\int_{-\infty}^{\infty} K_1(x-\xi-y)f(y)\,dy$$

is bounded. Hence, by dominated convergence,

(10·12) $$\int_{-\infty}^{\infty} Q(\xi)\,d\xi \lim_{x\to\infty}\int_{-\infty}^{\infty} K(x-\xi-y)f(y)\,dy$$
$$= \lim_{x\to\infty}\int_{-\infty}^{\infty} Q(\xi)\,d\xi \int_{-\infty}^{\infty} K(x-\xi-y)f(y)\,dy$$
$$= \lim_{x\to\infty}\int_{-\infty}^{\infty} f(y)\,dy \int_{-\infty}^{\infty} K(x-\xi-y)Q(\xi)\,d\xi,$$

as the integral converges absolutely. Combining (10·10) and (10·12), we see that if

$$K_2(x) = \int_{-\infty}^{\infty} K(x-\xi)Q(\xi)\,d\xi,$$

(10·12) is true.

Lemma 7₅. *If $K(x)$ belongs to Σ, a sub-class of M_1, and $Q(x)$ belongs to L_1 (not necessarily M_1), (10·08) belongs to $\epsilon'(\Sigma)$.*

To begin with, (10·08) belongs to M_1, for it is continuous by a simple application of X_{18}, and

(10·13) $$\sum_{-\infty}^{\infty} \max_{n\leqslant x\leqslant n+1} \left|\int_{-\infty}^{\infty} K(x-\xi)Q(\xi)\,d\xi\right|$$
$$\leqslant \int_{-\infty}^{\infty} |Q(\xi)|\,d\xi \sum_{-\infty}^{\infty} \max_{n\leqslant x\leqslant n+1} |K(x-\xi)|$$
$$\leqslant 2\int_{-\infty}^{\infty} |Q(\xi)|\,d\xi \sum_{-\infty}^{\infty} \max_{n\leqslant x\leqslant n+1} |K(x)|$$

as in (10·02). Formula (10·10) remains *mutatis mutandis* valid, and instead of (10·11), we see that

(10·14) $$\left|\int_{-\infty}^{\infty} K_1(x-\xi-y)\,dg(y)\right|$$
$$\leqslant \sum_{-\infty}^{\infty}\int_{n}^{n+1} |K_1(x-\xi-y)|\,d\int_{0}^{y}|dg(z)|$$
$$\leqslant \limsup_{-\infty<n<\infty}\int_{n}^{n+1} |dg(y)| \sum_{-\infty}^{\infty} \max_{n\leqslant y\leqslant n+1} |K_1(x-\xi-y)|$$
$$\leqslant 2\limsup_{-\infty<n<\infty}\int_{n}^{n+1} |dg(y)| \sum_{-\infty}^{\infty} \max_{n\leqslant x\leqslant n+1} |K_1(x)|.$$

78 THE GENERAL TAUBERIAN THEOREM

Hence, by dominated convergence,

$$(10\cdot 15) \quad \int_{-\infty}^{\infty} Q(\xi)\,d\xi \lim_{x\to\infty}\int_{-\infty}^{\infty} K(x-\xi-y)\,dg(y)$$

$$= \lim_{x\to\infty}\int_{-\infty}^{\infty} Q(\xi)\,d\xi \int_{-\infty}^{\infty} K(x-\xi-y)\,dg(y).$$

This integral converges absolutely, and (10·15) becomes (see X_{38})

$$(10\cdot 16) \quad \lim_{x\to\infty}\int_{-\infty}^{\infty} dg(y)\int_{-\infty}^{\infty} K(x-\xi-y)\,Q(\xi)\,d\xi.$$

If we combine this with the modified form of (10·10) in which $f(y)\,dy$ is replaced by $dg(y)$, and put $K_2(x)$ for (10·08), we obtain (10·07).

Lemma 6_6. *If $K_2(x)$ belongs to L_1, if $\{K_1^{(n)}(x)\}$ is a sequence of functions belonging to Σ, and if*

$$\lim_{n\to\infty}\int_{-\infty}^{\infty} |K_2(x) - K_1^{(n)}(x)|\,dx = 0,$$

then $K_2(x)$ belongs to $\epsilon(\Sigma)$.

In the first place, if $f(x)$ is bounded,

$$(10\cdot 17) \quad \lim_{n\to\infty}\int_{-\infty}^{\infty} [K_2(x-y) - K_1^{(n)}(x-y)]f(y)\,dy = 0$$

uniformly in x, and

$$(10\cdot 18) \quad \lim_{n\to\infty}\int_{-\infty}^{\infty} [K_2(x) - K_1^{(n)}(x)]\,dx = 0.$$

From (10·17) it follows that

$$(10\cdot 19) \quad \overline{\lim_{x\to\infty}}\int_{-\infty}^{\infty} K_2(x-y)f(y)\,dy$$

$$= \lim_{n\to\infty}\overline{\lim_{x\to\infty}}\int_{-\infty}^{\infty} K_1^{(n)}(x-y)f(y)\,dy$$

$$= \lim_{n\to\infty} A \int_{-\infty}^{\infty} K_1^{(n)}(x)\,dx$$

$$= A \int_{-\infty}^{\infty} K_2(x)\,dx,$$

THE GENERAL TAUBERIAN THEOREM

and similarly, that

$$(10\cdot 20) \quad \lim_{x \to \infty} \int_{-\infty}^{\infty} K_2(x-y) f(y) \, dy = A \int_{-\infty}^{\infty} K_2(x) \, dx.$$

However, (10·19) and (10·20) together are equivalent to (10·04), and lemma 6_6 is established. Similarly, we have

Lemma 7₆. *If $K_2(x)$ belongs to M_1, if $\{K_1^{(n)}(x)\}$ is a sequence of functions belonging to Σ, and if*

$$(10\cdot 21) \quad \lim_{n \to \infty} \sum_{k=-\infty}^{\infty} \max_{k \leqslant x \leqslant k+1} |K_2(x) - K_1^{(n)}(x)| = 0,$$

then $K_2(x)$ belongs to $\epsilon'(\Sigma)$.

Here we follow the argument of (10·14). This shows that

$$\left| \int_{-\infty}^{\infty} K_1^{(n)}(x-y) \, dg(y) - \int_{-\infty}^{\infty} K_2(x-y) \, dg(y) \right|$$

$$\leqslant 2 \limsup \int_n^{n+1} |dg(y)| \sum_{-\infty}^{\infty} \max_{n \leqslant x \leqslant n+1} |K_1^{(n)}(x) - K_2(x)|.$$

By (10·21), if $\int_n^{n+1} |dg(y)|$ is bounded,

$$\int_{-\infty}^{\infty} K_2(x-y) \, dg(y) = \lim_{n \to \infty} \int_{-\infty}^{\infty} K_1^{(n)}(x-y) \, dg(y)$$

uniformly in x. Formula (10·18) holds as before, and an argument precisely like that of (10·19) shows that

$$\overline{\lim_{x \to \infty}} \int_{-\infty}^{\infty} K_2(x-y) \, dg(y) = \lim_{x \to \infty} \int_{-\infty}^{\infty} K_2(x-y) \, dg(y)$$

$$= A \int_{-\infty}^{\infty} K_2(x) \, dx.$$

Here as $n \to \infty$,

$$(10\cdot 22) \quad \lim_{x \to \infty} \int_{-\infty}^{\infty} K_1^{(n)}(x-y) \, dg(y) \to \lim_{x \to \infty} \int_{-\infty}^{\infty} K_2(x-y) \, dg(y),$$

and it also follows from (10·18) that

$$(10\cdot 23) \quad A \int_{-\infty}^{\infty} K_1^{(n)}(x) \, dx \to A \int_{-\infty}^{\infty} K_2(x) \, dx.$$

If we combine (10·22) and (10·23) we obtain our result.

§11. Lemmas Concerning Functions whose Fourier Transforms Vanish for Large Arguments.

Lemma 6₇. *Let $f(x)$ be a continuous function defined over $(-\infty, \infty)$, and vanishing over $(-\infty, -\pi + \epsilon)$ and $(\pi - \epsilon, \infty)$. Let*

(11·01) $\qquad g(u) = \dfrac{1}{\sqrt{2\pi}} \displaystyle\int_{-\infty}^{\infty} f(x) e^{-iux} dx.$

Then the three following statements are equivalent:

(1) $\displaystyle\sum_{-\infty}^{\infty} |g(n)| < \infty;$

(2) $\displaystyle\sum_{-\infty}^{\infty} \max_{n \leqslant u \leqslant n+1} |g(u)| < \infty;$

(3) $\displaystyle\int_{-\infty}^{\infty} |g(u)| \, du < \infty.$

It is of course obvious that (2) implies (1) and (3). We need therefore only show that (1) implies (2) and that (3) implies (2). Let

(11·015) $\phi(x) = 1 \ (|x| \leqslant \pi - \epsilon); \ = \dfrac{\pi - |x|}{\epsilon} (\pi - \epsilon \leqslant |x| \leqslant \pi);$
$\qquad\qquad = 0 \ (\pi \leqslant |x|).$

The Fourier transform of $\phi(x)$ is

$$\sqrt{\dfrac{2}{\pi}} \int_0^\infty \phi(x) \cos ux \, dx = \sqrt{\dfrac{2}{\pi}} \int_0^\infty \phi(x) \dfrac{d \sin ux}{u}$$

$$= -\sqrt{\dfrac{2}{\pi}} \int_0^\infty \dfrac{\sin ux}{u} \phi'(x) \, dx$$

$$= \sqrt{\dfrac{2}{\pi}} \int_{\pi-\epsilon}^{\pi} \dfrac{\sin ux}{u \epsilon} dx$$

$$= \sqrt{\dfrac{2}{\pi}} \dfrac{\cos(\pi - \epsilon)u - \cos \pi u}{u^2 \epsilon}.$$

Thus the Fourier transform of $\phi(x) e^{-ivx}$ is

(11·02) $\qquad \dfrac{1}{\sqrt{2\pi}} \displaystyle\int_{-\infty}^{\infty} \phi(x) e^{-i(u+v)x} dx$

$$= \sqrt{\dfrac{2}{\pi}} \dfrac{\cos(\pi - \epsilon)(u+v) - \cos \pi(u+v)}{(u+v)^2 \epsilon}.$$

THE GENERAL TAUBERIAN THEOREM

For our present purpose all that is important concerning (11·02) is that it is dominated by an expression

$$\frac{A}{B+(u+v)^2}.$$

From this it immediately results that the coefficients of the Fourier series of $\phi(x)e^{-iux}$ over $(-\pi, \pi)$ are dominated by an expression

$$\frac{A}{B+(u+n)^2},$$

where A and B are positive. Thus, by (11·01), since by (11·015)

$$\frac{1}{\sqrt{2\pi}}\int_{-\infty}^{\infty} f(x)\,e^{-iux}\,dx = \frac{1}{\sqrt{2\pi}}\int_{-\infty}^{\infty} f(x)\,\phi(x)\,e^{-iux}\,dx,$$

we see that positive A and B exist such that

$$g(u) \leqslant \sum_{-\infty}^{\infty} \frac{A\,|g(n)|}{B+(u+n)^2}.$$

It follows that there exist positive A and B such that

$$\max_{m \leqslant u \leqslant m+1} |g(u)| \leqslant \sum_{-\infty}^{\infty} \frac{A\,|g(n)|}{B+(m+n)^2}.$$

Hence

(11·03) $\displaystyle\sum_{-\infty}^{\infty} \max_{n \leqslant u \leqslant n+1} |g(u)| \leqslant \sum_{m=-\infty}^{\infty} \sum_{n=-\infty}^{\infty} \frac{A\,|g(n)|}{B+(m+n)^2}$

$$= \sum_{n=-\infty}^{\infty} |g(u)| \sum_{m_1=-\infty}^{\infty} \frac{A}{B+m_1^2} \quad [m_1 = m+n]$$

$$= \text{const.} \sum_{-\infty}^{\infty} |g(n)|.$$

Again,

$$g(u) = \frac{1}{\sqrt{2\pi}}\int_{-\infty}^{\infty} f(x)\,\phi(x)\,e^{-iux}\,dx$$

$$= \frac{1}{\pi}\int_{-\infty}^{\infty} f(x)\,dx \int_{-\infty}^{\infty} \frac{\cos(\pi-\epsilon)(u+v) - \cos\pi(u+v)}{(u+v)^2\,\epsilon} e^{ivx}\,dv$$

$$\leqslant \int_{-\infty}^{\infty} \frac{A}{B+(u-v)^2} |g(v)|\,dv.$$

82 THE GENERAL TAUBERIAN THEOREM

Hence there exist positive A and B such that

$$\max_{m \leqslant u \leqslant n+1} |g(u)| \leqslant \int_{-\infty}^{\infty} \frac{A}{B + (m-v)^2} |g(v)| \, dv,$$

and

$$(11\cdot 04) \quad \sum_{-\infty}^{\infty} \max_{n \leqslant u \leqslant n+1} |g(u)| \leqslant \sum_{-\infty}^{\infty} \int_{-\infty}^{\infty} \frac{A}{B + (m-v)^2} |g(v)| \, dv$$

$$= \int_{-\infty}^{\infty} |g(v)| \, dv \sum_{-\infty}^{\infty} \frac{A}{B + (m-v)^2}$$

$$\leqslant \text{const.} \int_{-\infty}^{\infty} |g(v)| \, dv.$$

Formulae (11·03) and (11·04) complete the proof of lemma 6_7.

Lemma 6_8. *If $f(x)$ belongs to L_1 and has the Fourier transform $g(u)$, then*

$$(11\cdot 05) \quad \frac{1}{\pi} \int_{-\infty}^{\infty} f(x - \xi) \frac{1 - \cos a\xi}{a\xi^2} \, d\xi$$

belongs to L_1 and has the Fourier transform

$$(11\cdot 06) \quad \left(1 - \frac{|u|}{a}\right) g(u) \quad (|u| \leqslant a); \quad 0 \quad (|u| > a).$$

Here the proof that (11·05) belongs to L_1 proceeds as in (10·09). Moreover,

$$\frac{1}{\sqrt{2\pi}} \int_{-\infty}^{\infty} e^{-iux} \frac{dx}{\pi} \int_{-\infty}^{\infty} f(x - \xi) \frac{1 - \cos a\xi}{a\xi^2} \, d\xi$$

$$= \frac{1}{\sqrt{2\pi}} \int_{-\infty}^{\infty} f(x) \, e^{-iux} \, dx \, \frac{1}{\pi} \int_{-\infty}^{\infty} \frac{1 - \cos a\xi}{a\xi^2} e^{-iu\xi} \, d\xi$$

by the inversion of the order of integration of an absolutely convergent integral, and, by (5·08), we obtain (11·06). If we replace (10·09) by (10·13), we get:

Lemma 7_8. *If $f(x)$ belongs to M_1, (11·05) belongs to M_1.*

Let us now prove:

Lemma 6_9. *If $f(x)$ belongs to L_1,*

$$(11\cdot 07) \quad \lim_{a \to \infty} \int_{-\infty}^{\infty} \left| f(x) - \frac{1}{\pi} \int_{-\infty}^{\infty} f(x - \xi) \frac{1 - \cos a\xi}{a\xi^2} \, d\xi \right| dx = 0.$$

THE GENERAL TAUBERIAN THEOREM

We shall first prove this in the particular case where $f(x)=1$ over (α, β) and 0 elsewhere. Here

$$\frac{1}{\pi}\int_{-\infty}^{\infty} f(x-\xi)\,\frac{1-\cos a\xi}{a\xi^2}\,d\xi = \frac{1}{\pi}\int_{x-\alpha}^{x-\beta}\frac{1-\cos a\xi}{a\xi^2}\,d\xi$$

$$= \frac{1}{\pi}\int_{a(x-\alpha)}^{a(x-\beta)}\frac{1-\cos y}{y^2}\,dy.$$

The integral tends boundedly to $\frac{1}{2}\operatorname{sgn}(x-\beta) - \frac{1}{2}\operatorname{sgn}(x-\alpha)$ over any finite range. Furthermore, if $A > |\alpha|$, $A > |\beta|$,

$$\left[\int_{2A}^{\infty} + \int_{-\infty}^{-2A}\right]\frac{dx}{\pi}\left|\int_{-\infty}^{\infty} f(x-\xi)\,\frac{1-\cos a\xi}{a\xi^2}\,d\xi\right|$$

$$= \left[\int_{2A}^{\infty} + \int_{-\infty}^{-2A}\right]\frac{dx}{\pi}\left|\int_{a(x-\alpha)}^{a(x-\beta)}\frac{1-\cos y}{y^2}\,dy\right|$$

$$< \int_{A}^{\infty}\frac{2aA}{a^2(y-A)^2}\,dy \leqslant \frac{2}{a}.$$

It follows that (11·07) holds for this $f(x)$, and hence for all step-functions. By X_{16}, if $f(x)$ belongs to L_1, and $\epsilon > 0$, there is a step-function $f_1(x)$ such that

(11·073) $$\int_{-\infty}^{\infty}|f(x)-f_1(x)|\,dx < \epsilon.$$

Hence

(11·077) $$\int_{-\infty}^{\infty}\left|\frac{1}{\pi}\int_{-\infty}^{\infty} f(x-\xi)\,\frac{1-\cos a\xi}{a\xi^2}\,d\xi\right.$$

$$\left. -\frac{1}{\pi}\int_{-\infty}^{\infty} f_1(x-\xi)\,\frac{1-\cos a\xi}{a\xi^2}\,d\xi\right|dx$$

$$\leqslant \int_{-\infty}^{\infty}|f(x)-f_1(x)|\,dx\,\frac{1}{\pi}\int_{-\infty}^{\infty}\frac{1-\cos a\xi}{a\xi^2}\,d\xi < \epsilon.$$

Thus since (11·07) holds for all step-functions such as f_1, by (11·073) and (11·077),

$$\varlimsup_{a\to\infty}\int_{-\infty}^{\infty}\left|f(x) - \frac{1}{\pi}\int_{-\infty}^{\infty} f(x-\xi)\,\frac{1-\cos a\xi}{a\xi^2}\,d\xi\right|dx < 2\epsilon,$$

which is impossible unless (11·07) is true for $f(x)$.

84 THE GENERAL TAUBERIAN THEOREM

Another lemma of the same type is:

Lemma 7_9. *If $f(x)$ belongs to M_1,*

(11·08)
$$\lim_{a \to \infty} \sum_{n=-\infty}^{\infty} \max_{n \le x \le n+1} \left| f(x) - \frac{1}{\pi} \int_{-\infty}^{\infty} f(x-\xi) \frac{1-\cos a\xi}{a\xi^2} d\xi \right| = 0.$$

To begin with, let us suppose that $f(x) = {}_A f(x)$. In X_{55}, let (a, b) be $(-2A, 2A)$. Let

$$K(x, y, r) = \frac{1-r}{\pi} \frac{1 - \cos \frac{x-y}{1-r}}{(x-y)^2}.$$

Formula (4·12) is satisfied if x is interior to $(-2A, 2A)$, as

$$\lim_{r \to 1} \frac{1-r}{\pi} \int_{-2A}^{2A} \frac{1 - \cos \frac{x-y}{1-r}}{(x-y)^2} dy = \lim_{r \to 1} \frac{1}{\pi} \int_{-\frac{2A-x}{1-r}}^{\frac{2A-x}{1-r}} \frac{1 - \cos u}{u^2} du = 1.$$

Again K is non-negative, and tends uniformly to 0 for $|x-y| \ge \epsilon$ as $r \to 1$. Thus by X_{56}, since $f(x)$ is continuous by hypothesis,

(11·085) $$f(x) = \lim_{r \to 1} \frac{1-r}{\pi} \int_{-2A}^{2A} f(\xi) \frac{1 - \cos \frac{x-\xi}{1-r}}{(x-\xi)^2} d\xi$$

$$= \lim_{a \to \infty} \frac{1}{\pi} \int_{-\infty}^{\infty} f(x-\xi) \frac{1-\cos a\xi}{a\xi^2} d\xi$$

uniformly over any range interior to $(-2A, 2A)$. Furthermore, if x is outside $\left(-\frac{3A}{2}, \frac{3A}{2}\right)$, and ξ lies interior to $(-A, A)$, we have

$$\frac{1 - \cos a(x-\xi)}{a(x-\xi)^2} \le \frac{2}{a(|x|-A)^2} \le \frac{8}{aA^2},$$

and hence if $|x| > \frac{3A}{2}$ and $f(x) = {}_A f(x)$,

$$\left| \frac{1}{\pi} \int_{-\infty}^{\infty} f(x-\xi) \frac{1-\cos a\xi}{a\xi^2} d\xi \right| \le \frac{2}{\pi} \frac{\int_{-\infty}^{\infty} |f(\xi)| d\xi}{a(|x|-A)^2}.$$

THE GENERAL TAUBERIAN THEOREM

From this it follows that

$$(11\cdot09) \quad \lim_{a\to\infty} \left\{ \sum_{n=-\infty}^{-\left[\frac{3A}{2}+1\right]} + \sum_{n=\left[\frac{3A}{2}+1\right]}^{\infty} \right\}$$

$$\max_{n\leqslant x\leqslant n+1} \left| f(x) - \frac{1}{\pi}\int_{-\infty}^{\infty} f(x-\xi) \frac{1-\cos a\xi}{a\xi^2} d\xi \right|$$

$$= \lim_{a\to\infty} \left\{ \sum_{n=-\infty}^{-\left[\frac{3A}{2}+1\right]} + \sum_{n=\left[\frac{3A}{2}+1\right]}^{\infty} \right\}$$

$$\max_{n\leqslant x\leqslant n+1} \left| \frac{1}{\pi}\int_{-\infty}^{\infty} f(x-\xi) \frac{1-\cos a\xi}{a\xi^2} d\xi \right|$$

$$\leqslant \frac{2}{\pi}\int_{-\infty}^{\infty} |f(\xi)|\, d\xi \lim_{a\to\infty} \frac{1}{a} \left\{ \sum_{n=-\infty}^{-\left[\frac{3A}{2}+1\right]} + \sum_{n=\left[\frac{3A}{2}+1\right]}^{\infty} \right\} \frac{1}{(|x|-A)^2}$$

$$= \text{const.}\ \lim_{a\to\infty} \frac{1}{a} = 0.$$

On the other hand, it is a trivial consequence of (11·085) that

$$(11\cdot10) \quad \lim_{a\to\infty} \sum_{n=-\left[\frac{3A}{2}\right]}^{\left[\frac{3A}{2}\right]}$$

$$\max_{n\leqslant x\leqslant n+1} \left| f(x) - \frac{1}{\pi}\int_{-\infty}^{\infty} f(x-\xi) \frac{1-\cos a\xi}{a\xi^2} d\xi \right| = 0.$$

If we combine (11·09) and (11·10), (11·08) is the result.

Again, as in (10·13), if $f(x)$ is now *any* function belonging to M_1,

$$(11\cdot11) \quad \sum_{n=-\infty}^{\infty} \max_{n\leqslant x\leqslant n+1} \left| \frac{1}{\pi}\int_{-\infty}^{\infty} f(x-\xi) \frac{1-\cos a\xi}{a\xi^2} d\xi \right|$$

$$\leqslant \frac{2}{\pi}\int_{-\infty}^{\infty} \frac{1-\cos a\xi}{a\xi^2} d\xi \sum_{-\infty}^{\infty} \max_{n\leqslant x\leqslant n+1} |f(x)|.$$

Thus if $f(x)$ belongs to M_1, and $g(x, A)$ is defined by

$$g(x, A) = f(x)\left(1 - \frac{|x|}{A}\right) \quad (|x| < A);\quad = 0 \quad (|x| > A),$$

THE GENERAL TAUBERIAN THEOREM

since clearly

(11·12) $$\lim_{A\to\infty} \sum_{n=-\infty}^{\infty} \max_{n\leqslant x\leqslant n+1} |f(x)-g(x,A)| = 0,$$

(11·11) will yield us

(11·13) $$\lim_{A\to\infty} \sum_{n=-\infty}^{\infty} \max_{n\leqslant x\leqslant n+1} \left| \frac{1}{\pi}\int_{-\infty}^{\infty} f(x-\xi)\frac{1-\cos a\xi}{a\xi^2}d\xi \right.$$
$$\left. -\frac{1}{\pi}\int_{-\infty}^{\infty} g(x-\xi,A)\frac{1-\cos a\xi}{a\xi^2}d\xi \right| = 0$$

uniformly for all sufficiently large a. Since $g(x, A)$ belongs to a class of functions for which we have already proved (11·08) to hold, (11·12) and (11·13) complete the proof that it holds for every function $f(x)$ belonging to M_1.

§ 12. Lemmas on Absolutely Convergent Fourier Series.

Let $f(x)$ be a continuous function with Fourier series

(12·01) $$f(x) = \sum_{-\infty}^{\infty} c_n e^{inx}$$

over $(0, 2\pi)$. Let

$$\sum_{-\infty}^{\infty} |c_n| = C < \infty.$$

We shall say that $f(x)$ has an absolutely convergent Fourier series, shall write A for the class of all functions with absolutely convergent Fourier series, and shall put

$$C = A\{f\}.$$

It is obvious that:

Lemma 6₁₀. *If $f(x)$ and $g(x)$ both belong to A, so do $f(x) + g(x)$ and $f(x)g(x)$, and*

$$A\{f+g\} \leqslant A\{f\} + A\{g\};$$
(12·02) $$A\{fg\} \leqslant A\{f\}A\{g\}.$$

In the case of (12·02), let us note that if (12·01) converges absolutely, and if similarly

(12·03) $$g(x) = \sum_{-\infty}^{\infty} d_n e^{inx},$$

which also converges absolutely, then the double series we obtain by multiplying (12·01) and (12·03) converges absolutely, and may be rearranged in any order. Thus

$$f(x)\,g(x) = \sum_{m=-\infty}^{\infty} \sum_{n=-\infty}^{\infty} c_m d_n e^{i(m+n)x}$$

$$= \sum_{-\infty}^{\infty} e_n e^{inx},$$

where $\qquad e_n = \sum_{-\infty}^{\infty} c_m d_{n-m}.$

Thus $A\{fg\} = \sum_{n=-\infty}^{\infty} \left| \sum_{m=-\infty}^{\infty} c_m d_{n-m} \right| \leqslant \sum_{n=-\infty}^{\infty} \sum_{m=-\infty}^{\infty} |c_m d_{n-m}|$

$$= \sum_{-\infty}^{\infty} |c_m| \sum_{-\infty}^{\infty} |d_n| = A\{f\}\,A\{g\}.$$

We now turn to functions with absolutely convergent Fourier series, for which the term c_0 exceeds the sum of the moduli of all other terms. Such a function arises from an arbitrary function with absolutely convergent Fourier series on the addition of a sufficiently large constant. It can obviously have no zeros, as the constant term exceeds in modulus the sum of all other terms. As a corollary of lemma 6_{10}, we have:

Lemma 6_{11}. *If $f(x)$ belongs to A, and*

(12·04) $\qquad \left| \int_{-\pi}^{\pi} f(x)\,dx \right| > \pi A\{f\},$

then $1/f(x)$ belongs to A.

To prove this let us write the Fourier series of $f(x)$ as in (12·01). Then

$$1/f(x) = \frac{1}{c_0} \left(\frac{1}{1 + \left[\sum_1^{\infty} + \sum_{-\infty}^{-1} \right] \dfrac{c_n}{c_0} e^{inx}} \right)$$

$$= \frac{1}{c_0} \left\{ 1 - \left[\sum_1^{\infty} + \sum_{-\infty}^{-1} \right] \frac{c_n}{c_0} e^{inx} + \left\{ \left[\sum_1^{\infty} + \sum_{-\infty}^{-1} \right] \frac{c_n}{c_0} e^{inx} \right\}^2 - \ldots \right\},$$

THE GENERAL TAUBERIAN THEOREM

so that by lemma 6_{10},

$$A\{1/f\} \leq \frac{1}{|c_0|}\left\{1 + \left[\sum_1^\infty + \sum_{-\infty}^{-1}\right]\frac{|c_n|}{|c_0|} + \left\{\left[\sum_1^\infty + \sum_{-\infty}^{-1}\right]\frac{|c_n|}{|c_0|}\right\}^2 + \cdots\right\}$$

$$= \frac{1}{|c_0|}\frac{1}{1 - \left[\sum_1^\infty + \sum_{-\infty}^{-1}\right]\frac{|c_n|}{|c_0|}} = \frac{1}{|c_0| - \left[\sum_1^\infty + \sum_{-\infty}^{-1}\right]|c_n|}$$

$$= \frac{1}{2|c_0| - A\{f\}} = \frac{\pi}{\left|\int_{-\pi}^{\pi} f(x)\,dx\right| - \pi A\{f\}}.$$

Here all the series are convergent geometrical progressions or are dominated by them.

Lemma 6_{12}. *If $f(x)$ belongs to A, and $g_{abcd}(x)$ is defined by*

$$g_{abcd}(x) = 0 \quad (-\pi - \epsilon \leq x \leq a); \quad = \frac{x-a}{b-a} \quad (a \leq x \leq b);$$

$$= 1 \quad (b \leq x \leq c); \quad = \frac{d-x}{d-c} \quad (c \leq x \leq d);$$

$$= 0 \quad (d \leq x \leq \pi - \epsilon),$$

then $f(x) g_{abcd}(x)$ belongs to A if $a < b < c < d$.

The graph of $g_{abcd}(x)$ has the following form:

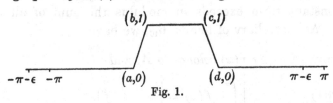

Fig. 1.

As will be seen, it consists entirely of segments of straight lines.

Lemma 6_{12} follows from 6_{10} and the fact that the nth Fourier coefficient of $f(x)$ is of the order of magnitude of n^{-2}.

Lemma 6_{13}. *If $f(x)$ belongs to A, $\eta > 0$, and $f(x_0) = 0$, then we may choose ϵ so small that*

$$A\{f(x) g_{x_0-2\epsilon,\, x_0-\epsilon,\, x_0+\epsilon,\, x_0+2\epsilon}(x)\} < \eta.$$

There is manifestly no restriction in taking x_0 to be 0. By (12·01),

(12·05) $$\sum_{-\infty}^{\infty} c_n = 0.$$

THE GENERAL TAUBERIAN THEOREM

Let
$$g_{-2\epsilon,-\epsilon,\epsilon,2\epsilon}(x) = \sum_{-\infty}^{\infty} d_n e^{inx}.$$

If we represent $g_{-2\epsilon,-\epsilon,\epsilon,2\epsilon}(x)$ by a Fourier integral, assuming it to vanish outside $(-\pi, \pi)$, its Fourier transform will be

(12·055)
$$\sqrt{\frac{2}{\pi}} \frac{\cos u\epsilon - \cos 2u\epsilon}{u^2 \epsilon}.$$

Similarly the Fourier transform of $g_{-2\epsilon,-\epsilon,\epsilon,2\epsilon}(x)(e^{imx}-1)$ will be

(12·06)
$$\sqrt{\frac{2}{\pi}} \left\{ \frac{\cos(u-m)\epsilon - \cos 2(u-m)\epsilon}{(u-m)^2 \epsilon} - \frac{\cos u\epsilon - \cos 2u\epsilon}{u^2 \epsilon} \right\}.$$

Clearly the integral of the modulus of (12·055) over $(-\infty, \infty)$ will be finite, while, by X_{18}, the integral of the modulus of (12·06) over $(-\infty, \infty)$ tends to 0 with ϵ, as

$$\int_{-\infty}^{\infty} \sqrt{\frac{2}{\pi}} \left| \frac{\cos(u-m)\epsilon - \cos 2(u-m)\epsilon}{(u-m)^2 \epsilon} - \frac{\cos u\epsilon - \cos 2u\epsilon}{u^2 \epsilon} \right| du$$

$$= \int_{-\infty}^{\infty} \sqrt{\frac{2}{\pi}} \left| \frac{\cos(v-m\epsilon) - \cos 2(v-m\epsilon)}{(v-m\epsilon)^2} - \frac{\cos v - \cos 2v}{v^2} \right| dv.$$

Thus by (11·03) in the proof of lemma 6_7, it follows that

$$\sum_{-\infty}^{\infty} |d_n| < \text{const.},$$

where this constant is independent of ϵ, and that

$$\lim_{\epsilon \to 0} \sum_{-\infty}^{\infty} |d_{n-m} - d_n| = 0.$$

Let us put
$$f(x) g_{-2\epsilon,-\epsilon,\epsilon,2\epsilon}(x) = \sum_{-\infty}^{\infty} a_n e^{inx}.$$

Let N be so large that

$$\sum_{N}^{\infty} |c_n| + \sum_{-\infty}^{-N} |c_n| < \eta.$$

Let ϵ then be so small that

$$\sum_{-\infty}^{\infty} |d_{n-m} - d_n| < \eta$$

THE GENERAL TAUBERIAN THEOREM

for $n \leqslant N$. Then

$$\sum_{-\infty}^{\infty} |a_n| = \sum_{n=-\infty}^{\infty} \left| \sum_{m=-\infty}^{\infty} d_{n-m} c_m \right|$$

$$\leqslant \text{const.} \, \eta + \sum_{n=-\infty}^{\infty} \left| \sum_{m=-N}^{N} d_{n-m} c_m \right|$$

$$\leqslant \text{const.} \, \eta + \sum_{n=-\infty}^{\infty} \left| d_n \sum_{m=-N}^{N} c_m \right| + \sum_{n=-\infty}^{\infty} \sum_{m=-N}^{N} |d_{n-m} - d_n| c_m$$

$$\leqslant \text{const.} \, \eta + \text{const.} \, \eta + \text{const.} \, \eta = \text{const.} \, \eta.$$

[In the second term, we make use of the fact that

$$\left| \sum_{m=-N}^{N} c_m \right| = \left| \sum_{-\infty}^{\infty} c_m - \sum_{-\infty}^{-N-1} c_m - \sum_{N+1}^{\infty} c_m \right| \leqslant |0 + \eta| = \eta,$$

which follows from (12·05).]

Hence $\qquad \lim\limits_{\epsilon \to 0} \sum\limits_{-\infty}^{\infty} |a_n| = 0.$

Lemma 6₁₄. *If $f(x)$ belongs to A and $f(x_0) \neq 0$, then there is a function $g(x)$ coinciding with $f(x)$ in some neighbourhood of x_0, belonging to A, and such that $1/g(x)$ belongs to A.*

There is no real restriction in taking x_0 to be 0. Let us define the function $g_{abcd}(x)$ as in lemma 6₁₂, and let us put

(12·07) $\qquad g(x) = f(0) + g_{-2\epsilon, -\epsilon, \epsilon, 2\epsilon}(x)(f(x) - f(0)).$

In lemma 6₁₃ we have shown that we may make

$$A \{g_{-2\epsilon, -\epsilon, \epsilon, 2\epsilon}(x)(f(x) - f(0))\}$$

less than η, by taking ϵ small enough. If the A of a function is less than η, the same is true of the modulus of each Fourier coefficient. Thus, by (12·07),

$$\left| \int_{-\pi}^{\pi} g(x) \, dx \right| > 2\pi f(0) - 2\pi \eta,$$

while, by another application of lemma 6₁₃,

$$\pi A \{g\} < \pi f(0) + \pi \eta.$$

Thus, in case $\qquad \eta < f(0)/3,$

(12·04) will be true, we may apply lemma 6₁₁, and lemma 6₁₄ will be established.

THE GENERAL TAUBERIAN THEOREM 91

Lemma 6$_{15}$. *If $f(x)$ has the property that whenever x_0 lies in $(-\pi, \pi)$, including the end points, there is an interval*

$$(x_0 - \epsilon, x_0 + \epsilon) \quad (\epsilon > 0)$$

and a function $g(x)$ belonging to A, such that over $(x_0 - \epsilon, x_0 + \epsilon)$,

$$g(x) = f(x),$$

then $f(x)$ belongs to A.

By the Heine-Borel theorem we can cover the interval $(-\pi, \pi)$ by a finite number of overlapping intervals $(x_0 - \epsilon, x_0 + \epsilon)$. Let these intervals be $(a_1, b_1), \ldots, (a_N, b_N)$, and let

$$a_1 < b_N - 2\pi < a_2 < b_1 < a_3 < b_2 < a_4 < b_3 < \ldots$$
$$< a_{N-1} < b_{N-2} < a_N < b_{N-1} < a_1 + 2\pi.$$

Let $g_k(x)$ be the function $g(x)$ coinciding with $f(x)$ over (a_k, b_k). Then

$$f(x) = \sum_1^N f(x)\, g_{a_k,\, b_{k-1},\, a_{k+1},\, b_k}(x)$$
$$= \sum_1^N g_k(x)\, g_{a_k,\, b_{k-1},\, a_{k+1},\, b_k}(x).$$

Here we put

$$2\pi + a_0 = a_N, \quad 2\pi + b_0 = b_N, \quad a_{N+1} = 2\pi + a_1, \quad b_{N+1} = 2\pi + b_1,$$

and suppose all our functions to be of period 2π. Our lemma follows from lemmas 6$_{10}$ and 6$_{12}$.

Lemma 6$_{16}$. *If $f(x)$ belongs to A, and does not vanish anywhere in $(-\pi, \pi)$, $1/f(x)$ belongs to A.*

This is an immediate consequence of lemmas 6$_{14}$ and 6$_{15}$.

Lemma 6$_{17}$. *If $f(x)$ belongs to A, and does not vanish in the neighbourhood of a and b, there is a function $g(x)$, not vanishing outside (a, b), belonging to A, and coinciding with $f(x)$ over (a, b).*

We may assume without essential restriction that

$$-\pi < a < b < \pi.$$

Let δ be so small that if x lies in $(a - \delta, a)$,

(12·08) $\quad |f(x) - f(a)| < \tfrac{1}{2}|f(a)|,$

and that if x lies in $(b, b + \delta)$,

(12·085) $\quad |f(x) - f(b)| < \tfrac{1}{2}|f(b)|.$

92 THE GENERAL TAUBERIAN THEOREM

This will always be possible, as $f(x)$ does not vanish in the neighbourhood of a and of b. Let

$$g(x) = \exp\left\{i + \frac{x+\pi}{a+\pi-\delta}\left\{\log\left\{\frac{|f(a)|}{f(a)}\right\} - i\right\}\right\}; \quad (-\pi < x \leqslant a - \delta)$$

$$g(x) = f(x)\frac{x-a+\delta}{\delta} + \frac{f(a)}{|f(a)|}\frac{a-x}{\delta}; \quad (a-\delta < x \leqslant a)$$

$$g(x) = f(x); \quad (a < x \leqslant b)$$

$$g(x) = f(x)\frac{b-x+\delta}{\delta} + \frac{f(b)}{|f(b)|}\frac{x-b}{\delta}; \quad (b < x \leqslant b+\delta)$$

$$g(x) = \exp\left\{\log\left\{\frac{|f(b)|}{f(b)}\right\}\right.$$
$$\left. + \frac{x-b-\delta}{\pi-b-\delta}\left\{i - \log\left\{\frac{|f(b)|}{f(b)}\right\}\right\}\right\}. \quad (b+\delta < x \leqslant \pi)$$

Clearly $g(x)$ cannot vanish over

$$(-\pi < x \leqslant a - \delta) \quad \text{or} \quad (b < x \leqslant b + \delta).$$

If $g(x)$ were to vanish over $(a-\delta, a)$, we should necessarily have the argument of $f(x)$ π plus the argument $f(a)$. This would contradict (12·08). Similarly, in view of (12·085), $g(x)$ cannot vanish over $(b, b+\delta)$.

The function $g(x)$ is the sum of a function with a bounded differential coefficient—and hence with an absolutely convergent Fourier series—with the function

$$f(x) g_{a-\delta, a, b, b+\delta}(x),$$

which belongs to A by lemmas 6_{12} and 6_{10}. Hence by lemma 6_{10}, $g(x)$ belongs to A.

Lemma 6_{18}. *If $f(x)$ and $g(x)$ belong to A, and the zeros of $g(x)$ are all interior points of a finite number of intervals over which $f(x)$ is everywhere 0, then $f(x)/g(x)$ belongs to A.* Here we have $f(x)/g(x)$ to be 0 when $f(x) = 0$.

We may write (on consideration of lemma 6_{17})

(12·09) $\quad f(x)/g(x) = f(x)/h(x),$

where $h(x)$ has no zeros over $(-\pi, \pi)$, for we may modify $g(x)$ in a finite number of stages so that it shall not vanish in turn

THE GENERAL TAUBERIAN THEOREM 93

in each of the intervals over which $f(x)$ is everywhere 0, without changing it outside these intervals, and may thus eventually obtain $h(x)$. By lemma 6_{16}, $1/h(x)$ belongs to A, and thus by (12·09) and lemma 6_{10}, $f(x)/g(x)$ belongs to A.

It is perhaps worth while to indicate the significance of the class A when we write our Fourier series in the real sine-cosine form. Of course, if

$$f(x) = \sum_{-\infty}^{\infty} c_n e^{inx},$$

the absolute convergence of the series for $f(x)$ is independent of x, and we may therefore define the class of functions with absolute convergent Fourier series as the class for which $\sum_{-\infty}^{\infty} |c_n|$ converges. In case we put

$$f(x) = a_0/2 + \sum_{1}^{\infty} (a_n \cos nx + b_n \sin nx),$$

we have for the sum of the absolute values of the terms of this series

$$(12·10) \quad |a_0/2| + \sum_{1}^{\infty} |a_n \cos nx + b_n \sin nx|$$

$$= c_0 + \sum_{1}^{\infty} |c_n e^{inx} + c_{-n} e^{-inx}| \leqslant \sum_{-\infty}^{\infty} |c_n|.$$

This may however converge without $\sum_{-\infty}^{\infty} |c_n|$ converging, as in the case where $x = 0$, $c_n = -c_{-n}$.

However, we have

$$|c_n e^{inx} + c_{-n} e^{-inx}| = |c_{-n} + c_n e^{2inx}|$$
$$\geqslant |c_n| |\sin 2nx|$$
$$\geqslant |c_n| \sin^2 2nx$$

and also
$$\geqslant |c_{-n}| \sin^2 2nx.$$

Hence $\quad |c_n e^{inx} + c_{-n} e^{-inx}| \geqslant \dfrac{|c_n| + |c_{-n}|}{2} \sin^2 2nx$

$$= \dfrac{|c_n| + |c_{-n}|}{4} (1 - \cos 4nx).$$

94 THE GENERAL TAUBERIAN THEOREM

Now let the series (12·10) converge to a limit $\leqslant L$ over a set of points S of measure greater than 0. Let $S(x)$ be the function which is 1 over S and 0 elsewhere. By the principle of bounded convergence, we may integrate

$$S(x)\left\{|a_0/2| + \sum_1^\infty |a_n \cos nx + b_n \sin nx|\right\}$$

term by term between finite limits, and hence may integrate

$$\Sigma S(x) \frac{|c_n| + |c_{-n}|}{4}(1 - \cos 4nx)$$

between finite limits. Thus

$$\sum_1^\infty \frac{|c_n| + |c_{-n}|}{4}\left\{m(S) - \int_{-\pi}^{\pi} S(x) \cos 4nx\, dx\right\}$$

converges. Thus, by the Riemann-Lebesgue theorem,

$$|c_0| + \sum_1^\infty \{|c_n| + |c_{-n}|\}\{1 + O(1)\}$$

converges, and hence

(12·11) $$\sum_{-\infty}^{\infty} |c_n|$$

converges. Thus if (12·10) converges over more than a null set, (12·11) converges. In particular, if the absolute convergence of a Fourier series is taken to mean its absolute convergence everywhere, it is equivalent to the convergence of (12·11)*.

§ 13. The Proof of the General Tauberian Theorem.

Let us return to the classes $\epsilon(\Sigma)$ and $\epsilon'(\Sigma)$. We have

Lemma 6₁₉. *The function $K(x)$ of the class L_1 belongs to a class $\epsilon(\Sigma)$ when and only when*

(13·01) $$\frac{1}{\pi}\int_{-\infty}^{\infty} K(x - \xi)\frac{1 - \cos a\xi}{a\xi^2}\, d\xi$$

belongs to $\epsilon(\Sigma)$ for all a.

First let K belong to $\epsilon(\Sigma)$. Then by lemma 6₅, (13·01) belongs to $\epsilon(\epsilon(\Sigma))$, and hence by lemma 6₁, to $\epsilon(\Sigma)$. Conversely, let

---

* Lemma 6₁₈ is due to A. Denjoy, "Sur l'absolue convergence des séries trigonométriques," *C. R.* 155 (1912), 135–136; N. Lusin, "Sur l'absolue convergence des séries trigonométriques," *ibid.* 580–582.

THE GENERAL TAUBERIAN THEOREM

(13·01) belong to $\epsilon(\Sigma)$ for all a. Then by lemmas 6_9 and 6_6, $K(x)$ belongs to $\epsilon(\epsilon(\Sigma))$, and hence to $\epsilon(\Sigma)$. A similar argument depending on lemmas 7_1, 7_5, 7_6, and 7_9 leads to:

Lemma 7_{19}. *The function $K(x)$ of M_1 belongs to a class $\epsilon'(\Sigma)$ when and only when (13·01) belongs to $\epsilon'(\Sigma)$ for all a.*

Lemma 6_{20} (7_{20}). *If $K(x)$ belongs to Σ and has the Fourier transform $k(u)$, which does not vanish over the closed interval (α, β), while $K_2(x)$ is a function belonging to L_1, and hence by lemma 6_7 to M_1, which has a Fourier transform vanishing over (α, β), then $K_2(x)$ belongs to $\epsilon(\Sigma)$ [or $\epsilon'(\Sigma)$].*

To begin with, if $K_2(x)$ has the Fourier transform $k_2(u)$, it follows, from lemmas 6_7, 6_8 [and 7_8], and 6_{18}, that

$$(13\cdot02) \quad k_3(u) = k_2(u)\left(1 - \frac{|u|}{a}\right)\bigg/k(u) \ (|u| < a); \ k_3(u) = 0 \ (|u| \geqslant a)$$

belongs to A for sufficiently great a, the class A now being interpreted to refer to functions with a period $2B$ ($B > a$), instead of 2π. It follows from 6_7 that $k_3(u)$ has a Fourier transform belonging to L_1. However, $k_3(u)$ is a function belonging to L_2, as it is bounded and differs from 0 only over a finite range. Hence its Fourier transform belongs to L_2 as well as L_1. Let the Fourier transform of $k_3(u)$ be $K_3(-x)$. Then, by the Plancherel theorem,

$$k_3(u) = \lim_{A \to \infty} \frac{1}{\sqrt{2\pi}} \int_{-A}^{A} K_3(x) e^{-iux} dx.$$

However, as $K_3(x)$ belongs to L_1,

$$(13\cdot03) \quad \frac{1}{\sqrt{2\pi}} \int_{-\infty}^{\infty} K_3(x) e^{-iux} dx$$

exists for every u, and is almost everywhere equal to $k_3(u)$, by X_{41}. Thus $k_3(u)$ is the Fourier transform of a function $K_3(x)$ belonging to L_1.

The function

$$(13\cdot04) \quad \frac{1}{\sqrt{2\pi}} \int_{-\infty}^{\infty} K(x-y) K_3(y) \, dy$$

96 THE GENERAL TAUBERIAN THEOREM

will belong to $\epsilon(\Sigma)[\epsilon'(\Sigma)]$ by lemma $6_5[7_5]$. Its Fourier transform will be

$$\frac{1}{2\pi}\int_{-\infty}^{\infty} e^{-iux}\,dx \int_{-\infty}^{\infty} K(x-y)K_3(y)\,dy$$

$$=\frac{1}{2\pi}\int_{-\infty}^{\infty} K_3(y)e^{-iuy}\,dy \int_{-\infty}^{\infty} K(x-y)e^{-iu(x-y)}\,dx$$

$$=k_3(u)k(u).$$

Thus, by (13·02), the function $k_4(u)$, defined by

$$k_4(u) = k_2(u)\left(1-\frac{|u|}{a}\right) \quad (|u|<a); \quad k_4(u)=0 \quad (|u|\geqslant a),$$

has (13·04) as its Fourier transform. Hence, by lemma 6_8,

$$\frac{1}{\pi}\int_{-\infty}^{\infty} K_2(x-\xi)\frac{1-\cos a\xi}{a\xi^2}\,d\xi = \frac{1}{\sqrt{2\pi}}\int_{-\infty}^{\infty} K(x-y)K_3(y)\,dy,$$

where $K_3(y)$ belongs to L_1. The lemma now follows from lemmas 6_{19} [or 7_{19}] and 6_5 [or 7_5].

We are thus in a position to prove theorems 6 and 7. As the proof is entirely parallel in the two cases, we shall only consider theorem 6. If $a<\infty$, by the Heine-Borel theorem, we may divide the interval $(-2a, 2a)$ into a finite number of overlapping intervals I_n, in each of which the Fourier transform of some single function $K_1(x)$ of Σ has no zeros. Thus by lemma 6_{20} *any* function of L_1 with a Fourier transform differing from 0 only over an interval interior to one of these intervals belongs to $\epsilon(\Sigma)$. Now let $K_2(x)$ be any function of class L_1. Let its Fourier transform be $k_2(u)$. As in lemma 6_{15}, we may write

$$k_2(u)\left(1-\frac{|u|}{a}\right) = \sum_{1}^{N} k_2(u)\left(1-\frac{|u|}{a}\right)g_{a_k, b_{k-1}, a_{k+1}, b_k}(x)$$

over $(-a, a)$ where each term of the sum differs from 0 only over an interval interior to an interval I_n. Each term is moreover a function with an absolutely convergent Fourier series, by lemmas 6_{10} and 6_{12}, and hence by lemma 6_7 has a Fourier transform belonging to L_1. By the argument of (13·03) and (13·04), it is therefore the Fourier transform of a function of class L_1, and

THE GENERAL TAUBERIAN THEOREM 97

the Fourier transform of $k_2(u)\left(1 - \dfrac{|u|}{a}\right)$ is the sum of these transforms. Thus

(13·05) $\quad \dfrac{1}{\pi}\displaystyle\int_{-\infty}^{\infty} K_2(x-\xi)\dfrac{1-\cos a\xi}{a\xi^2}d\xi$

is the sum of a finite number of functions belonging to $\epsilon(\Sigma)$, and hence belongs to $\epsilon(\Sigma)$. Here we again introduce the argument of (13·03) and (13·04). Then, by lemma 6_{19}, $K_2(x)$ belongs to $\epsilon(\Sigma)$, and the proof is complete.

§14. The Closure of the Translations of a Function of L_1.

We have incidentally proved:

THEOREM 8. *If $K_1(x)$ belongs to L_1, and*

(14·01) $\quad \dfrac{1}{\sqrt{2\pi}}\displaystyle\int_{-\infty}^{\infty} K_1(x)\, e^{iux}\, dx \neq 0$

for all real values of u, if $K_2(x)$ belongs to L_1, and if $\epsilon > 0$, then there exists a function $K_3(x)$ belonging to L_1, vanishing for arguments of large modulus and such that

(14·02) $\quad \displaystyle\int_{-\infty}^{\infty}\left|K_2(x) - \int_{-\infty}^{\infty} K_1(x-\xi)K_3(\xi)d\xi\right|dx < \epsilon.$

This will be clear if we reflect on our proof of theorem 6, for we showed that $K_2(x)$ could be approximated in the sense of (14·02) by (13·05), which we had shown to be representable in the form

$$\int_{-\infty}^{\infty} K_1(x-\xi)K_3(\xi)d\xi,$$

where K_3 vanishes for arguments of large modulus and belongs to L_1. We now wish to prove that if $K_1(x)$ and $K_3(x)$ both belong to L_1, then

(14·03) $\quad \displaystyle\lim_{\eta\to 0}\int_{-\infty}^{\infty}\left|\sum_{-\infty}^{\infty} K_1(x-n\eta)\int_{n\eta}^{(n+1)\eta} K_3(\xi)d\xi\right.$

$\left. - \displaystyle\int_{-\infty}^{\infty} K_1(x-\xi)K_3(\xi)d\xi\right|dx = 0.$

98 THE GENERAL TAUBERIAN THEOREM

The expression of which we take the limit in (14·03) does not exceed

$$\int_{-\infty}^{\infty}\left|\sum_{-\infty}^{\infty}\int_{n\eta}^{(n+1)\eta} K_3(\xi)\left[K_1(x-n\eta)-K_1(x-\xi)\right]d\xi\right|dx$$

$$\leqslant \sum_{-\infty}^{\infty}\int_{n\eta}^{(n+1)\eta}|K_3(\xi)|\,d\xi\int_{-\infty}^{\infty}|K_1(x-n\eta)-K_1(x-\xi)|\,dx$$

$$\leqslant \int_{-\infty}^{\infty}|K_3(\xi)|\,d\xi\,\max_{\xi\leqslant\eta}\int_{-\infty}^{\infty}|K_1(x+\xi)-K_1(x)|\,dx.$$

From X_{18} we get (14·03).

We have thus established the direct part of

THEOREM 9. *If $K_1(x)$ belongs to L_1, and for all real u*

(14·01) $\qquad \dfrac{1}{\sqrt{2\pi}}\int_{-\infty}^{\infty} K_1(x)\,e^{iux}\,dx \neq 0,$

then if $K_2(x)$ belongs to L_1, and $\epsilon > 0$, there exists an integer N, together with a set of real numbers Λ_n and complex numbers A_n ($n = 1, 2, ..., N$), such that

(14·04) $\qquad \int_{-\infty}^{\infty}\left|K_2(x)-\sum_1^N A_n K_1(x-\Lambda_n)\right|dx < \epsilon.$

Conversely, let K_1 belong to L_1, and let it be possible, whenever K_2 belongs to L_1 and $\epsilon > 0$, to find quantities N, A_n and Λ_n as above, for which (14·04) holds. Then (14·01) is true.

As for the converse part of theorem 9, let

(14·05) $\qquad \dfrac{1}{\sqrt{2\pi}}\int_{-\infty}^{\infty} K_1(x)\,e^{iu_0 x}\,dx = 0,$

and let $\qquad \dfrac{1}{\sqrt{2\pi}}\int_{-\infty}^{\infty} K_2(x)\,e^{iu_0 x}\,dx \neq 0.$

Then $\int_{-\infty}^{\infty}\left|K_2(x)-\sum_1^N A_n K_1(x-\Lambda_n)\right|dx$

$$\geqslant \left|\int_{-\infty}^{\infty}\left[K_2(x)-\sum_1^N A_n K_1(x-\Lambda_n)\right]e^{iu_0 x}\,dx\right|$$

$$=\left|\int_{-\infty}^{\infty} K_2(x)\,e^{iu_0 x}\,dx\right|,$$

so that (14·04) is impossible for arbitrarily small values of ϵ.

THE GENERAL TAUBERIAN THEOREM 99

A more general theorem which we may establish in the same way is:

THEOREM 10. *If Σ is a class of functions belonging to L_1, then the two following statements are equivalent:*

(1) *There is no real number u_0 for which* (14·05) *holds whenever $K_1(x)$ belongs to Σ.*

(2) *If $K_2(x)$ belongs to L_1 and $\epsilon > 0$, there exist integers $N_1, \ldots, N_k$, functions $Q_1(x), \ldots, Q_k(x)$ belonging to Σ, real numbers $\Lambda_{n,j}$ for $j \leqslant k$, $n \leqslant N_j$, and complex numbers $A_{n,j}$ for the same range of suffixes, such that*

$$(14\cdot06) \quad \int_{-\infty}^{\infty} \left| K_2(x) - \sum_{j=1}^{k} \sum_{n=1}^{N_k} A_{n,j} Q_j(x - \Lambda_{n,j}) \right| dx < \epsilon.$$

We shall not give the proof of this theorem in full. It depends on covering the interval $(-\infty, \infty)$ by a denumerable set of overlapping intervals I in each of which the Fourier transform of some one function of Σ does not vanish, in such a way that any interval $(-A, A)$ is covered by a finite number of intervals I. We then show much as in theorem 9 that if $K_2(x)$ is a function of L_1 whose Fourier transform vanishes except over an interval interior to an interval I, and $K_1(x)$ is a function of Σ whose Fourier transform has no zeros on I, then an integer N, a set of real numbers Λ_n and complex numbers A_n ($n = 1, 2, \ldots, N$) may be found such that (14·04) holds. The proof proceeds much as in the case of theorem 9.

Once (14·04) is established for functions $K_2(x)$ with Fourier transforms vanishing except over individual intervals of I, (14·06) is an immediate corollary for functions $K_2(x)$ with Fourier transforms vanishing except over some finite interval, for, as we have said, every finite interval is the sum of a finite number of intervals of I. By means of theorem 8 it is not difficult to complete the proof of the direct part of theorem 10. The inverse part proceeds exactly as in theorem 9.

A complete formal proof is left as an exercise to the reader.

§15. The Closure of the Translations of a Function of L_2.

THEOREM 11. *If $K_1(x)$ belongs to L_2, and almost everywhere on $-\infty < u < \infty$,*

$$(15\cdot01) \qquad \underset{A\to\infty}{\text{l.i.m.}} \frac{1}{\sqrt{2\pi}} \int_{-A}^{A} K_1(x) e^{iux} dx \neq 0,$$

then if $K_2(x)$ belongs to L_2, and $\epsilon > 0$, there exists an integer N, together with a set of real numbers Λ_n and complex numbers $A_n (n = 1, 2, \ldots, N)$, such that

$$(15\cdot02) \qquad \int_{-\infty}^{\infty} \left| K_2(x) - \sum_{1}^{N} A_n K_1(x - \Lambda_n) \right|^2 dx < \epsilon.$$

Conversely, let K_1 belong to L_2, and let it be possible, whenever K_2 belongs to L_2, and $\epsilon > 0$, to find quantities N, A_n, and Λ_n as above, for which (15·02) holds. Then (15·01) is true almost everywhere.

The word "almost" distinguishes this theorem from theorem 9. It is extremely natural to find this distinction between the L_2 theorem and the corresponding L_1 theorem, as the Fourier transform of a function of L_1 is defined everywhere, while that of a function of L_2 is only defined "almost everywhere."

In the proof of theorem 11, we have the advantage that the theory of the Fourier transform is completely symmetrical in L_2, which is not true of any other Lebesgue class. The class of Fourier transforms of functions of L_2 is L_2 itself, so that it is possible to translate theorem 11 into the following equivalent theorem:

THEOREM 12. *If $k_1(u)$ belongs to L_2, and only has a set of zeros of zero measure, then if $k_2(u)$ belongs to L_2, and $\epsilon > 0$, there exists an integer N, together with a set of real numbers Λ_n and complex numbers $A_n (n = 1, 2, \ldots, N)$, such that*

$$(15\cdot03) \qquad \int_{-\infty}^{\infty} \left| k_2(u) - k_1(u) \sum_{1}^{N} A_n e^{i\Lambda_n u} \right|^2 du < \epsilon.$$

Conversely, let k_1 belong to L_2, and let it be possible, whenever k_2 belongs to L_2, and $\epsilon > 0$, to find quantities N, A_n, and Λ_n as

THE GENERAL TAUBERIAN THEOREM 101

above, for which (15·03) *holds. Then* $k_1(u)$ *has at most a null set of zeros.* In this translation we have made use of the fact that the integral of the square of the modulus of a function is equal to the integral of the square of the modulus of its Fourier transform.

Now, as to the proof: let $\eta > 0$, and let A be so large that

(15·035) $\qquad \left[\int_{-\infty}^{-A} + \int_{A}^{\infty}\right] |k_2(u)|^2 du < \eta.$

Let $\qquad f(u) = k_2(u)/k_1(u),$

and let B be so large that

(15·04) $\qquad \int_{-\infty}^{\infty} |f_B(u) k_1(u) - k_2(u)|^2 du < \eta.$

We can find such a B, since $|f_B(u) k_1(u) - k_2(u)|^2$ is a sequence decreasing everywhere and almost everywhere tending to 0, so that by monotone convergence,

$$\lim_{B \to \infty} \int_{-\infty}^{\infty} |f_B(u) k_1(u) - k_2(u)|^2 du = 0.$$

Let it be noted that we have here made use of the fact that $k_1(u)$ has at most a null set of zeros.

It will result from (15·035) and the definition of $f_B(u)$ that

(15·05) $\qquad \left[\int_{-\infty}^{-A} + \int_{A}^{\infty}\right] |f_B(u) k_1(u)|^2 du < \eta.$

Combining (15·05) and (15·04), we have, by the Minkowski inequality,

(15·06) $\qquad \int_{-\infty}^{\infty} |k_2(u) - {}_A f_B(u) k_1(u)|^2 du < 4\eta.$

Let $g(u, C)$ be equal to ${}_A f_B(u)$ over $(-C, C)$, where $C > A$, and let it have the period $2C$. We have

$$\int_{-\infty}^{\infty} |{}_A f_B(u) k_1(u) - g(u, C) k_1(u)|^2 du$$

$$= \sum_{1}^{\infty} \left\{ \int_{-C}^{C} |{}_A f_B(u) k_1(u + 2nC)|^2 du \right.$$

$$\left. + \int_{-C}^{C} |{}_A f_B(u) k_1(u - 2nC)|^2 du \right\}$$

$$\leq B^2 \int_{C}^{\infty} |k_1(u)|^2 du + B^2 \int_{-\infty}^{-C} |k_1(u)|^2 du.$$

102 THE GENERAL TAUBERIAN THEOREM

Let C be so large—as is clearly possible—that

$$\int_{-\infty}^{\infty} |{}_Af_B(u)k_1(u) - g(u,C)k_1(u)|^2 \, du < \eta.$$

Then, by (15·06) and the Minkowski inequality,

(15·07) $\quad \int_{-\infty}^{\infty} |k_2(u) - g(u,C)k_1(u)|^2 \, du < 9\eta.$

Let $\qquad g(u,C) \sim \sum_{-\infty}^{\infty} a_n e^{\frac{\pi i n u}{C}}.$

Clearly $\qquad a_n = \dfrac{1}{2C} \int_{-C}^{C} g(u,C) e^{-\frac{\pi i n u}{C}} \, du.$

Accordingly

$$\left| \sum_{-N}^{N} \left(1 - \frac{|n|}{N}\right) a_n e^{\frac{\pi i n u}{C}} \right|$$

$$= \left| \frac{1}{2C} \int_{-C}^{C} g(v,C) \, dv \sum_{-N}^{N} \left(1 - \frac{|n|}{N}\right) e^{\frac{\pi i n (v-u)}{C}} \right|$$

$$\leq B \int_{-C}^{C} \left| \sum_{-N}^{N} \left(1 - \frac{|n|}{N}\right) e^{\frac{\pi i n (v-u)}{C}} \right| \, dv.$$

However, we may readily verify that

$$\sum_{-N}^{N} \left(1 - \frac{|n|}{N}\right) e^{inx} = \frac{\sin^2 \frac{Nx}{2}}{\sin^2 \frac{x}{2}} \geq 0.$$

This verification may, for example, be made by mathematical induction. Hence

(15·08)
$$\left| \sum_{-N}^{N} \left(1 - \frac{|n|}{N}\right) a_n e^{\frac{\pi i n u}{C}} \right| \leq B \int_{-C}^{C} \left[\sum_{-N}^{N} \left(1 - \frac{|n|}{N}\right) e^{\frac{\pi i n (v-u)}{C}} \right] dv$$
$$= 2BC.$$

Again,

$$\int_{-C}^{C} \left| g(u,C) - \sum_{-N}^{N} \left(1 - \frac{|n|}{N}\right) a_n e^{\frac{\pi i n u}{C}} \right|^2 du$$

$$= 2C \left\{ \sum_{-\infty}^{\infty} |a_n|^2 - \sum_{-N}^{N} \left(1 - \frac{n^2}{N^2}\right) |a_n|^2 \right\},$$

THE GENERAL TAUBERIAN THEOREM 103

which tends to zero with increasing N. Thus we can find some sequence of values N_k of N for which

$$\left| g(u, C) - \sum_{-N_k}^{N_k} \left(1 - \frac{|n|}{N}\right) a_n e^{\frac{\pi i n u}{C}} \right| \to 0$$

for almost all values of u. It follows by dominated convergence [in view of (15·08)] that we may choose N_k so large that

$$\int_{-\infty}^{\infty} \left| g(u, C) k_1(u) - \sum_{-N_k}^{N_k} \left(1 - \frac{|n|}{N_k}\right) a_n e^{\frac{\pi i n u}{C}} k_1(u) \right|^2 du < \eta,$$

and hence, by the Minkowski inequality, because of (15·07), that

$$\int_{-\infty}^{\infty} \left| k_2(u) - \sum_{-N_k}^{N_k} \left(1 - \frac{|n|}{N_k}\right) a_n e^{\frac{\pi i n u}{C}} k_1(u) \right|^2 du < 25\eta.$$

Thus if $\eta = \epsilon/25$, we have established the existence of an inequality such as (15·03). This completes the direct part of the proof of theorem 12, and hence of theorem 11.

As to the inverse part, if $k_1(u)$ vanishes over more than a null set, let $k_2(u) = 1$ over a part of this set of positive measure, and 0 elsewhere. Then

$$\int_{-\infty}^{\infty} \left| k_2(u) - k_1(u) \sum_{1}^{N} A_n e^{i\Lambda_n u} \right|^2 du$$

is at least equal to the measure of the set where $k_2(u) = 1$ and $k_1(u) = 0$. Thus (15·03) is impossible.

CHAPTER III

SPECIAL TAUBERIAN THEOREMS

§ 16. The Abel-Tauber Theorem.

The first theorem to be known as Tauberian was proved in 1897 by A. Tauber*. It asserts that if $\sum_{0}^{\infty} a_n x^n = f(x)$ converges for $0 \leqslant x < 1$, if $a_n = o(1/n)$, and if $f(x)$ tends to s as $x \to 1$ from below, then $\sum_{0}^{\infty} a_n = s$. It is therefore a conditional converse of Abel's theorem, which asserts that if $\sum_{0}^{\infty} a_n$ converges, then

$$\lim_{x \to 1-0} \sum_{0}^{\infty} a_n x^n = \sum_{0}^{\infty} a_n.$$

The unconditional converse of Abel's theorem is false. For example, $\sum_{0}^{\infty} (-1)^n$ diverges, yet

$$\sum_{0}^{\infty} (-x)^n = \frac{1}{1+x}$$

for $|x| < 1$, and this tends to $\frac{1}{2}$ as $x \to 1$. However, Tauber's conclusion holds under assumptions concerning a_n less restrictive than those of Tauber. The fundamental theorem here is due to Littlewood†. Littlewood's theorem is:

THEOREM 13. *Let $\sum_{0}^{\infty} a_n x^n$ converge to $f(x)$ for $|x| < 1$. Let*

(16·01) $$\lim_{x \to 1-0} f(x) = s$$

as x tends to 1 along the real axis. Let $n|a_n| < K < \infty$. Then

(16·02) $$\sum_{0}^{\infty} a_n = s.$$

* A. Tauber, "Ein Satz aus der Theorie der unendlichen Reihen." *Monatshefte f. Math.* 8 (1897), 273–277.

† J. E. Littlewood, "On the converse of Abel's theorem on power series." *Proc. Lond. Math. Soc.* (2), 9 (1910), 434–444.

SPECIAL TAUBERIAN THEOREMS

To begin with, let us put
$$s(x) = \sum_0^{[x]} a_n.$$
We have

(16·03) $\quad |s(x) - f(e^{-\frac{1}{x}})| = \left| \sum_0^{[x]} a_n (1 - e^{-\frac{n}{x}}) - \sum_{[x]+1}^{\infty} a_n e^{-\frac{n}{x}} \right|$

$$\leqslant \sum_0^{[x]} \frac{K}{n} \frac{n}{x} + \sum_{[x]+1}^{\infty} \frac{K}{n} e^{-\frac{n}{x}}$$

$$\leqslant 2K + K \int_{[x]}^{\infty} e^{-\frac{u}{x}} \frac{du}{u}$$

$$\leqslant 3K + K \int_1^{\infty} e^{-u} \frac{du}{u} = \text{const.}$$

However, $f(e^{-\frac{1}{x}})$ is bounded for $0 \leqslant x < \infty$. Hence $s(x)$ is bounded.

Now, $\quad f(e^{-x}) = \sum_0^{\infty} a_n e^{-nx} = \int_{-0}^{\infty} e^{-ux} ds(u)$

$$= \int_0^{\infty} xe^{-ux} s(u) du$$

by X_{36}. Hence

(16·04) $\quad s = \lim_{x \to 0} \int_0^{\infty} xe^{-ux} s(u) du$

$$= \lim_{\xi \to \infty} \int_{-\infty}^{\infty} e^{-\xi} e^{-e^{\eta-\xi}} s(e^\eta) e^\eta d\eta.$$

Let us put $\quad K_1(\xi) = e^{-\xi} e^{-e^{-\xi}}.$
We have
$$\int_{-\infty}^{\infty} K_1(\xi) d\xi = \int_{-\infty}^{\infty} e^{-\xi} e^{-e^{-\xi}} d\xi = \int_0^{\infty} e^{-x} dx = 1.$$
Formula (16·04) becomes
$$\lim_{\xi \to \infty} \int_{-\infty}^{\infty} K_1(\xi - \eta) s(e^\eta) d\eta = s \int_{-\infty}^{\infty} K_1(\xi) d\xi,$$
where $s(e^{-\eta})$ is bounded. Furthermore,
$$\frac{1}{\sqrt{2\pi}} \int_{-\infty}^{\infty} K_1(\xi) e^{-iu\xi} d\xi = \frac{1}{\sqrt{2\pi}} \int_0^{\infty} x^{iu} e^{-x} dx$$
$$= \frac{1}{\sqrt{2\pi}} \Gamma(1+iu) \neq 0.$$

106 SPECIAL TAUBERIAN THEOREMS

Thus the hypothesis of theorem 4 is satisfied, and if we take
(16·045) $K_2(\xi) = 0$ $(\xi < 0);$ $K_2(\xi) = e^{-\xi}$ $(\xi > 0),$
we get, by (10·04),

$$(16·05) \quad s = s \int_0^\infty e^{-\xi} d\xi = s \int_{-\infty}^\infty K_2(\xi) d\xi$$

$$= \lim_{\xi \to \infty} \int_{-\infty}^\infty K_2(\xi - \eta) s(e^\eta) d\eta$$

$$= \lim_{\xi \to \infty} \int_{-\infty}^\xi e^{\eta - \xi} s(e^\eta) d\eta$$

$$= \lim_{x \to \infty} \frac{1}{x} \int_0^x s(y) dy.$$

It follows that, if $\lambda > 0$,
(16·06)

$$s = \frac{(1+\lambda)s - s}{\lambda} = \lim_{x \to \infty} \frac{1}{\lambda x} \left[\int_0^{(1+\lambda)x} s(y) dy - \int_0^x s(y) dy \right]$$

$$= \lim_{x \to \infty} \frac{1}{\lambda x} \int_x^{(1+\lambda)x} s(y) dy$$

$$= \lim_{x \to \infty} \left\{ s(x) + \frac{1}{\lambda x} \int_x^{(1+\lambda)x} \{s(y) - s(x)\} dy \right\}.$$

Now $\left| \frac{1}{\lambda x} \int_x^{(1+\lambda)x} \{s(y) - s(x)\} dy \right| \leq \frac{1}{\lambda x} \int_x^{(1+\lambda)x} dy \sum_{[x]+1}^{[y]} \frac{K}{n}$

$$\leq \sum_{[x]+1}^{[(1+\lambda)x]} \frac{K}{[x]} \leq \frac{[\lambda x] K}{[x]} < 2\lambda K$$

for sufficiently large values of x. Hence, by (16·06),

$$\varlimsup_{x \to \infty} |s(x) - s| < 2\lambda K,$$

and since λ is any positive quantity,
(16·07) $\lim_{x \to \infty} |s(x) - s| = 0.$

This is, however, merely another mode of writing (16·02). We have thus completed the proof of theorem 13*.

An extremely simple proof for this theorem has been given by Karamata†. It proceeds as follows: if (16·04) holds, and $K_3(\xi)$ is any function of L_1 for which we can find for any $\epsilon > 0$ a polynomial

$$(16·08) \quad P_n(\xi) = \sum_1^n a_k e^{\log k \cdot \xi} e^{-e^{\log k \cdot \xi}}$$

* Cf. Schmidt 1; Wiener 2. † Karamata 1.

such that
$$\int_{-\infty}^{\infty} |K_3(x) - P_n(\xi)| \, d\xi < \epsilon,$$
then, by lemma 6_6,

(16·09) $\quad \lim\limits_{\xi \to \infty} \int_{-\infty}^{\infty} K_3(\xi - \eta) s(e^\eta) \, d\eta = s \int_{-\infty}^{\infty} K_3(\xi) \, d\xi.$

Now, we know by the famous theorem of Weierstrass on polynomial approximation that if $f(x)/x$ is any continuous function of x over $(0, 1)$, we may uniformly approximate to $f(x)$ over $(0, 1)$ by a polynomial of the form

$$\sum_{1}^{n} k a_k x^{k-1}.$$

Thus, if $P_n(x)$ is defined as in (16·08),

$$\int_{-\infty}^{\infty} |e^{-\xi} f(e^{-e^{-\xi}}) - P_n(\xi)| \, d\xi$$

$$= \int_{0}^{\infty} \left| f(e^{-x}) - \sum_{1}^{n} k a_k e^{-kx} \right| dx$$

$$= \int_{0}^{1} \left| \frac{f(y)}{y} - \sum_{1}^{n} k a_k y^{k-1} \right| dy$$

$$\leqslant \max_{0 < y < 1} \left| \frac{f(y)}{y} - \sum_{1}^{n} k a_k y^{k-1} \right|.$$

It follows that if

(16·095) $\quad\quad K_3(\xi) = e^{-\xi} f(e^{-e^{-\xi}}),$

where $f(x)/x$ is continuous over the closed interval $(0, 1)$, (16·09) will hold. Now let us take

$$K_3(\xi) = 0 \quad (\xi < -\epsilon); \quad K_3(\xi) = \frac{\xi + \epsilon}{\epsilon} \quad (-\epsilon \leqslant \xi < 0);$$

$$K_3(\xi) = e^{-\xi} \quad (\xi > 0).$$

If we define $f(x)$ by (16·095), clearly $\dfrac{f(x)}{x}$ will be continuous at any interior point of $(0, 1)$. However, over the interval $(0, e^{-e^\epsilon})$, $\dfrac{f(x)}{x}$ will be identically zero, and in the neighbourhood of 1, $f(x)/x$ will be identically $1/x$. Thus (16·09) is established.

108 SPECIAL TAUBERIAN THEOREMS

Furthermore, if we take as $K_2(\xi)$ the function defined in (16·045), clearly

$$\lim_{\epsilon \to 0} \int_{-\infty}^{\infty} |K_3(\xi) - K_2(\xi)| \, d\xi = 0.$$

Thus, by another application of lemma 6_6, (16·05) will follow, and we may complete the proof of theorem 13 as above.

Let it be noted that Karamata's proof avoids any reference to the Fourier transform of $K_2(\xi)$, because the particular form of $K_2(\xi)$ makes a reference to the Weierstrass theorem possible. However, we have seen in theorems 11 and 12 that in the general case, it is not possible to discuss problems of closure or completeness analogous to those of the Weierstrass theorem without entering a region where we find necessary and sufficient conditions depending on the distribution of the zeros of the Fourier transform of a certain kernel function. Thus the Karamata methods, though far simpler in the cases where they are natural to apply, cannot be expected to have the scope of the methods of the last chapter.

In the hypothesis of theorem 13, we may replace* the condition of the boundedness of na_n by the apparently weaker condition

(16·10) $\qquad\qquad na_n > -K.$

We shall then have

$$f(e^{-\frac{1}{x}}) - f(e^{-\frac{2}{x}}) = \sum_0^\infty a_n (e^{-\frac{n}{x}} - e^{-\frac{2n}{x}})$$

$$= \sum_0^{[x]} a_n (e^{-\frac{n}{x}} - e^{-\frac{2n}{x}}) + \sum_{[x]+1}^{[2x]} a_n (e^{-\frac{n}{x}} - e^{-\frac{2n}{x}})$$

$$+ \sum_{[2x]+1}^{\infty} a_n (e^{-\frac{n}{x}} - e^{-\frac{2n}{x}})$$

$$\geqslant -\sum_0^{[x]} \frac{K}{n}\frac{n}{x} + \sum_{[x]+1}^{[2x]} a_n (e^{-2} - e^{-4}) - \sum_{[x]+1}^{[2x]} \frac{K}{n}(e^{-1} - 2e^{-2} + e^{-4})$$

$$- \sum_{[2x]+1}^{\infty} \frac{K}{n} e^{-\frac{n}{x}} - e^{-\frac{2n}{x}}.$$

* Cf. Landau 1.

SPECIAL TAUBERIAN THEOREMS

Here we have made use of the fact that $e^{-x} - e^{-2x}$ has its maximum at $x = \log 2$, and is decreasing over $(1, 2)$.

It follows from this and (16·01) that

$$\varlimsup_{x \to \infty} \sum_{[x]+1}^{[2x]} a_n \leqslant \frac{1}{e^{-2} - e^{-4}} \left\{ K + 2K(e^{-1} - 2e^{-2} + e^{-4}) + K \int_2^\infty \frac{e^{-u}}{u} du \right\}.$$

Here the last integral enters as in (16·03). Combining this with (16·10), we see that there exists a Q independent of N, such that

$$\left| \sum_{N+1}^{2N} a_n \right| < Q.$$

It follows that

(16·11) $\quad \sum_{[x]+1}^{\infty} a_n e^{-\frac{n}{x}} = \sum_{[x]+1}^{[2x]} a_n e^{-\frac{n}{x}} + \sum_{[2x]+1}^{[4x]} a_n e^{-\frac{n}{x}} + \ldots$

$\leqslant Q(e^{-1} + e^{-2} + e^{-4} + \ldots)$

$\quad + \sum_{[x]+1}^{[2x]} a_n(e^{-\frac{n}{x}} - e^{-1}) + \sum_{[2x]+1}^{[4x]} (e^{-\frac{n}{x}} - e^{-2}) + \ldots$

$\leqslant \frac{Q}{1+e} + e^{-1} \sum_{[x]+1}^{[2x]} \frac{K}{n}\left(\frac{n}{x} - 1\right) + e^{-2} \sum_{[2x]+1}^{[4x]} \frac{K}{n}\left(\frac{n}{x} - 2\right) + \ldots$

$\leqslant \frac{Q}{1+e} + K\left(\frac{1}{e} + \frac{2}{e^2} + \frac{2^2}{e^4} + \frac{2^3}{e^8} + \ldots\right) < \text{const.}$

Similarly,

(16·12) $\quad \sum_{[x]+1}^{\infty} a_n e^{-\frac{n}{x}} \geqslant -Q(e^{-2} + e^{-4} + \ldots)$

$\quad + \sum_{[x]+1}^{[2x]} a_n(e^{-\frac{n}{x}} - e^{-2}) + \sum_{[2x]+1}^{[4x]} (e^{-\frac{n}{x}} - e^{-4}) + \ldots$

$> \text{const.}$

Hence $\quad \left| \sum_{[x]+1}^{\infty} a_n e^{-\frac{n}{x}} \right| < \text{const.}$

Since $f(x)$ is bounded,

(16·13) $\quad \left| \sum_0^{[x]} a_n e^{-\frac{n}{x}} \right| < \text{const.}$

Hence, if $\xi = 2x$,

$$\left| \sum_{0}^{[x]} a_n e^{-\frac{n}{2x}} \right| = \left| \sum_{0}^{[\xi/2]} a_n e^{-\frac{n}{\xi}} \right|$$

$$= \left| \sum_{0}^{[\xi]} a_n e^{-\frac{n}{\xi}} - \sum_{[x]+1}^{[2x]} a_n e^{-\frac{n}{2x}} \right|$$

$$\leqslant \text{const.} + \left| \sum_{[x]+1}^{[2x]} a_n e^{-\frac{n}{2x}} \right|,$$

and hence, by an argument similar to that of (16·11) and (16·12),

$$\left| \sum_{0}^{[x]} a_n e^{-\frac{n}{2x}} \right| < \text{const.}$$

It follows that

(16·14) $\qquad \left| \sum_{0}^{[x]} a_n \left(3e^{-\frac{n}{2x}} - 2e^{-\frac{n}{x}} \right) \right| < \text{const.}$

Here we should notice that over $(0, [x])$,

$$1 + \frac{n}{2x} \geqslant 3e^{-\frac{n}{2x}} - 2e^{-\frac{n}{x}} \geqslant 1.$$

From (16·10) and (16·13) it follows that

$$\sum_{0}^{[x]} a_n = \sum_{0}^{[x]} a_n e^{-\frac{n}{x}} + \sum_{0}^{[x]} a_n \left(1 - e^{-\frac{n}{x}} \right)$$

$$\geqslant \text{const.} - \frac{K}{x} \sum_{0}^{[x]} \frac{1}{n} \frac{n}{x}$$

$$\geqslant \text{const.}$$

Similarly, it follows from (16·10) and (16·14) that

$$\sum_{0}^{[x]} a_n = \sum_{0}^{[x]} a_n \left(3e^{-\frac{n}{2x}} - 2e^{-\frac{n}{x}} \right) + \sum_{0}^{[x]} a_n \left(1 + 2e^{-\frac{n}{x}} - 3e^{-\frac{n}{2x}} \right)$$

$$\leqslant \text{const.} + \frac{K}{2x} \sum_{0}^{[x]} \frac{1}{n} \frac{n}{x}$$

$$\leqslant \text{const.}$$

SPECIAL TAUBERIAN THEOREMS

Thus $s(x)$ is bounded, and we may proceed as before to (16·05) and (16·06). Thus

$$\varlimsup_{x\to\infty} s(x) \leqslant s - \lim_{x\to\infty} \frac{1}{\lambda x}\int_x^{(1+\lambda)x} \{s(y) - s(x)\}\,dy$$

$$\leqslant s + \frac{1}{\lambda x}\int_x^{(1+\lambda)x} dy \sum_{[x]+1}^{[y]} \frac{K}{n}$$

$$< s + 2\lambda K,$$

and since λ is arbitrary,

(16·15) $$\varlimsup_{x\to\infty} s(x) \leqslant s.$$

Again, we may write (16·06),

$$s = \lim_{x\to\infty}\left\{s(x(1+\lambda)) + \frac{1}{\lambda x}\int_x^{(1+\lambda)x}\{s(y) - s(x(1+\lambda))\}\,dy\right\},$$

so that

$$\varliminf_{x\to\infty} s(x) \geqslant s + \lim_{x\to\infty}\frac{1}{\lambda x}\int_x^{(1+\lambda)x}\{s(x(1+\lambda)) - s(y)\}\,dy$$

$$\geqslant s - \frac{1}{\lambda x}\int_x^{(1+\lambda)x} dy \sum_{[y]+1}^{[(1+\lambda)x]} \frac{K}{n}$$

$$> s - 2\lambda K,$$

and since λ is arbitrary,

(16·16) $$\varliminf_{x\to\infty} s(x) \geqslant s.$$

Combining (16·15) and (16·16), we obtain (16·07) or (16·02), and have established:

THEOREM 14. *Let* $\sum_0^\infty a_n x^n$ *converge to* $f(x)$ *for* $|x| < 1$. *Let*

(16·01) $$\lim_{x\to 1-0} f(x) = s.$$

Let

(16·10) $$na_n > -K.$$

Then

(16·02) $$\sum_0^\infty a_n = s,$$

§ 17. The Prime-Number Theorem as a Tauberian Theorem.

The present section and the three following will be devoted to the application of Tauberian theorems to the problem of the distribution of the primes. The theorem which we shall eventually prove is the famous theorem of Hadamard and de la Vallée Poussin*, and reads:

THEOREM 15. *If $\pi(u)$ is the number of primes not exceeding u, then*

(17·01) $$\pi(n) \sim \frac{n}{\log n}.$$

Here the symbol $A(n) \sim B(n)$ is taken to mean that as $n \to \infty$, $A(n)/B(n) \to 1$.

As the first step in the proof of theorem 15, we shall establish:

Lemma 15$_1$. $\pi(n) = o(n)$.

Let $p_1, p_2, \ldots$ be the (finite or infinite) sequence of prime numbers. Clearly if we take any sequence of $p_1 p_2 \ldots p_n$ consecutive integers,

(17·02) $$\prod_1^n \left(1 - \frac{1}{p_k}\right) p_1 p_2 \ldots p_n$$

of them will not be divisible by any p_k ($k = 1, 2, \ldots, n$). We may see this if we subtract from $p_1 p_2 \ldots p_k$ the number of integers of the interval divisible by one p_k, add again the number divisible by two distinct p_k's, which have been subtracted twice, subtract the number divisible by three p_k's, and so on. This will give us

$$p_1 p_2 \ldots p_n \left(1 - \Sigma \frac{1}{p_k} + \Sigma \frac{1}{p_j p_k} - \Sigma \frac{1}{p_j p_k p_l} + \ldots \right),$$

each sum being taken over all combinations of distinct indices. This however is identically (17·02). Thus ultimately

$$\pi(n) \Big/ n \leqslant \prod_1^n \left(1 - \frac{1}{p_k}\right),$$

and if the product $\prod_1^\infty (1 - 1/p_k)$ diverges to 0, lemma 15$_1$ is established.

* Cf. Landau 2; A. E. Ingham (forthcoming Cambridge Tract on the Distribution of Prime Numbers).

SPECIAL TAUBERIAN THEOREMS

On the other hand, let the infinite product converge. Then

(17·03) $$\Sigma 1/p_k < \infty.$$

This is however impossible if $\pi(n)/n$ does not tend to 0. For let $\pi(n) \geqslant An$ for an infinite sequence of values of n. Then, for an infinite sequence of values of n,

$$p_n \leqslant n/A,$$

and $$\sum_{[n/2]}^{n} 1/p_k \geqslant A/2,$$

which contradicts (17·03) if $A > 0$.

Let us now introduce the number-theoretic functions,

(17·04) $$\varpi(u) = \pi(u) + \tfrac{1}{2}\pi(u^{\frac{1}{2}}) + \tfrac{1}{3}\pi(u^{\frac{1}{3}}) + \ldots,$$

and $\Lambda(n)$, defined by

$\Lambda(p^k) = \log p$ if p is a prime and k is a positive integer;

$\Lambda(n) = 0$ if n is not of the form p^k.

These functions are considerably more regular than $\pi(n)$, and it is of advantage to transform (17·01) into a form concerning them.

We may easily see that $\pi(u) = 0$ if $u < 2$, so that (17·04) is really a terminating series, with the last non-zero term determined by the highest integral root of u to equal or exceed 2. Thus we have

(17·05) $$|\varpi(u) - \pi(u)| = \left| \tfrac{1}{2}\pi(u^{\frac{1}{2}}) + \ldots + \frac{\pi\left(\exp \dfrac{\log u}{[\log u/\log 2 + 1]}\right)}{[\log u/\log 2] + 1} \right|$$

$$\leqslant u^{\frac{1}{2}} [\log u/\log 2]$$

$$= O(u^{\frac{1}{2}} \log u).$$

From (17·05) and lemma 15_1, it follows that

(17·06) $$\varpi(u) = o(u).$$

The result to be established in theorem 15 becomes

(17·07) $$\varpi(n) \sim \frac{n}{\log n},$$

SPECIAL TAUBERIAN THEOREMS

as $n^{\frac{1}{2}} \log n$ is of smaller order than $n/\log n$. This may again be written

(17·08) $$\lim_{N \to \infty} \frac{\varpi(N) \log N}{N} = 1.$$

Now, by (17·06),

$$\lim_{N \to \infty} \frac{1}{N} \int_0^N \frac{\varpi(u)\, du}{u} = \lim_{N \to \infty} \frac{1}{N} \int_0^N o(1)\, du = o(1).$$

Thus we may write (17·08)

$$\lim_{N \to \infty} \left\{ \frac{\varpi(N) \log N}{N} - \frac{1}{N} \int_0^N \frac{\varpi(u)\, du}{u} \right\} = 1,$$

or, upon integrating by parts,

$$1 = \lim_{N \to \infty} \frac{1}{N} \int_0^N \log u\, d\varpi(u)$$
$$= \lim_{N \to \infty} \frac{1}{N} \sum_{n=1}^N \log n\, (\varpi(n+0) - \varpi(n-0)).$$

However, as $\varpi(u)$ is continuous except for a jump of $1/\nu$ at every point p^ν, where p is a prime, we have

$$\Lambda(n) = \log n\, (\varpi(n+0) - \varpi(n-0)).$$

Hence theorem 15 will be established if we can show that

(17·085) $$\lim_{N \to \infty} \frac{1}{N} \sum_{n=1}^N \Lambda(n) = 1.$$

Let us now consider series of the type

(17·09) $$\sum_1^\infty a_n \frac{x^n}{1-x^n}.$$

These series are known as Lambert series, after their eighteenth century discoverer*. Formally such a series is equivalent to

(17·10) $$\sum_{n=1}^\infty x^n \sum_{m/n} a_m,$$

the coefficient being the sum of all coefficients a_m for values of m that are factors of n. This is a result of the fact that

(17·11) $$\frac{x^m}{1-x^m} = x^m + x^{2m} + x^{3m} + \cdots.$$

* Cf. K. Knopp, *Unendliche Reihen*, § 58 C.

SPECIAL TAUBERIAN THEOREMS

In case the coefficients a_n are such that

$$\sum_{1}^{\infty} a_n x^n$$

converges for $|x| < 1$, it is easy to see that for $|x| < 1$, (17·09) will converge absolutely, and that the rearrangement which yields (17·10) will be legitimate over the same range. In particular,

$$\sum_{n=1}^{\infty} \Lambda(n) \frac{x^n}{1-x^n} = \sum_{1}^{\infty} x^n \sum_{m|n} \Lambda(m).$$

Now let $\qquad n = p_1^{k_1} p_2^{k_2} \ldots p_\nu^{k_\nu},$

where $p_1, \ldots, p_\nu$ are primes and $k_1, \ldots, k_\nu$ positive integers. The factors of n for which the function Λ will assume a value other than 0 will be

$$p_1, p_1^2, \ldots, p_1^{k_1}; \; p_2, p_2^2, \ldots, p_2^{k_2}; \; \ldots; \; p_\nu, p_\nu^2, \ldots, p_\nu^{k_\nu}.$$

The result of applying Λ to these factors will be

$$\log p_1, \log p_1, \ldots, \log p_1; \; \log p_2, \log p_2, \ldots, \log p_2; \; \ldots;$$
$$\log p_\nu, \log p_\nu, \ldots, \log p_\nu.$$

Thus
$$\sum_{m|n} \Lambda(m) = k_1 \log p_1 + k_2 \log p_2 + \ldots + k_\nu \log p_\nu = \log n.$$

Hence $\qquad \sum_{1}^{\infty} \Lambda(n) \dfrac{x^n}{1-x^n} = \sum_{1}^{\infty} x^m \log m.$

By Abel's lemma on series,

$$\sum_{1}^{\infty} x^m \log m = \sum_{1}^{\infty} \log m \frac{x^m - x^{m+1}}{1-x}$$

$$= \sum_{1}^{\infty} \frac{x^{m+1}}{1-x} [\log(m+1) - \log m]$$

$$= \frac{x}{1-x} \sum_{1}^{\infty} \log\left(1 + \frac{1}{m}\right) x^m$$

$$= \frac{x}{1-x} \sum_{1}^{\infty} \left[\frac{1}{m} + O\left(\frac{1}{m^2}\right)\right] x^m$$

$$= \frac{x}{1-x} \left[\log \frac{1}{1-x} + \sum_{1}^{\infty} O\left(\frac{1}{m^2}\right) x^m\right].$$

116 SPECIAL TAUBERIAN THEOREMS

Let us now put $x = e^{-\xi}$, and let us multiply by $-\xi$. We have

(17·12)
$$\sum_{1}^{\infty} \Lambda(n) \frac{\xi e^{-n\xi}}{e^{-n\xi}-1} = \frac{\xi e^{-\xi}}{1-e^{-\xi}}\left[\log(1-e^{-\xi}) - \sum_{1}^{\infty} O\left(\frac{1}{m^2}\right)e^{-m\xi}\right].$$

The formal derivatives of the series on the two sides are, respectively,

(17·13)
$$\sum_{1}^{\infty} \Lambda(n) \frac{d}{dn\xi}\left(\frac{n\xi e^{-n\xi}}{e^{-n\xi}-1}\right) = \sum_{1}^{\infty} \Lambda(n) \frac{(1-n\xi)e^{-n\xi}(e^{-n\xi}-1) + n\xi e^{-2n\xi}}{(e^{-n\xi}-1)^2}$$

$$= \sum_{1}^{\infty} \Lambda(n) \frac{e^{-2n\xi} - e^{-n\xi} + n\xi e^{-n\xi}}{(e^{-n\xi}-1)^2},$$

and

(17·14) $$\frac{e^{-\xi} - e^{-2\xi} - \xi e^{-\xi}}{(1-e^{-\xi})^2}\left[\log(1-e^{-\xi}) - \sum_{1}^{\infty} O\left(\frac{1}{m^2}\right)e^{-m\xi}\right]$$

$$+ \frac{\xi e^{-\xi}}{1-e^{-\xi}}\left[\frac{e^{-\xi}}{1-e^{-\xi}} - \sum_{1}^{\infty} O\left(\frac{1}{m}\right)e^{-m\xi}\right].$$

For $\xi > 0$, the series of (17·13) is dominated by an expression of the form

$$\text{const. } \sum_{1}^{\infty} \Lambda(n) n\xi e^{-n\xi},$$

and converges uniformly for $\xi > \epsilon$. Thus (17·13) may be integrated term by term, or what is the same, the left-hand series of (17·12) may be differentiated term by term. The same is clearly true of the right-hand series, as (17·14) converges uniformly for $\xi > \epsilon$. Hence

$$\sum_{1}^{\infty} \Lambda(n) \frac{d}{dn\xi} \frac{n\xi e^{-n\xi}}{e^{-n\xi}-1} = \sum_{1}^{\infty} \Lambda(n) \frac{e^{-2n\xi} - e^{-n\xi} + n\xi e^{-n\xi}}{(e^{-n\xi}-1)^2}$$

$$= \frac{e^{-\xi} - e^{-2\xi} - \xi e^{-\xi}}{(1-e^{-\xi})^2}\left[\log(1-e^{-\xi}) + \sum_{1}^{\infty} O\left(\frac{1}{m^2}\right)e^{-m\xi}\right]$$

$$+ \frac{\xi e^{-\xi}}{1-e^{-\xi}}\left[\frac{e^{-\xi}}{1-e^{-\xi}} + \sum_{1}^{\infty} O\left(\frac{1}{m}\right)e^{-m\xi}\right]$$

$$= O(1)[O(\log \xi) + O(1)] + [1 + O(\xi)]\left[\frac{1}{\xi} + O(1) + O(\log \xi)\right]$$

$$= 1/\xi + O(\log \xi)$$

SPECIAL TAUBERIAN THEOREMS

as $\xi \to 0$. It follows that

(17·15) $\quad \lim\limits_{\xi \to 0} \xi \sum\limits_1^\infty \Lambda(n) \dfrac{e^{-2n\xi} - e^{-n\xi} + n\xi e^{-n\xi}}{(e^{-n\xi} - 1)^2} = 1.$

Let us put $\quad g(y) = \sum\limits_{n=1}^{[e^y]} \dfrac{\Lambda(n)}{n}.$

Then (17·15) assumes the form

$$1 = \lim_{\xi \to 0} \int_0^\infty \eta\xi \frac{e^{-2\eta\xi} - e^{-\eta\xi} + \eta\xi e^{-\eta\xi}}{(e^{-\eta\xi} - 1)^2} \, dg(\log \eta)$$

$$= \lim_{x \to \infty} \int_{-\infty}^\infty \frac{e^{y-x}(e^{-2e^{y-x}} - e^{-e^{y-x}} + e^{y-x}e^{-e^{y-x}})}{(e^{-e^{y-x}} - 1)^2} \, dg(y).$$

This is of the form of (10·06), if we put $A = 1$ and

$$K_1(x) = \frac{e^{-x}(e^{-2e^{-x}} - e^{-e^{-x}} + e^{-x}e^{-e^{-x}})}{(e^{-e^{-x}} - 1)^2},$$

and remember that

$$\int_{-\infty}^\infty K_1(x) \, dx = \int_0^\infty \frac{e^{-2\xi} - e^{-\xi} + \xi e^{-\xi}}{(e^{-\xi} - 1)^2} \, d\xi$$

$$= \int_0^\infty \frac{d}{d\xi}\left(\frac{\xi e^{-\xi}}{e^{-\xi} - 1}\right)$$

$$= \lim_{\xi \to 0} \frac{\xi e^{-\xi}}{1 - e^{-\xi}} = 1.$$

The function $K_1(x)$ belongs to M_1, and is always positive. For this we have only to prove that if $\xi > 0$,

$$e^{-2\xi} - e^{-\xi} + \xi e^{-\xi} \geqslant 0,$$

or that $\quad e^{-\xi} > 1 - \xi,$

which is a well-known inequality. Furthermore, $g(y)$ is monotone increasing. Thus

(17·16) $\quad 1 = \lim\limits_{x \to \infty} \displaystyle\int_{-\infty}^\infty K_1(x-y) \, dg(y)$

$\geqslant \overline{\lim\limits_{x \to \infty}} \displaystyle\int_x^{x+1} K(x-y) \, dg(y)$

$\geqslant \overline{\lim\limits_{x \to \infty}} \min\limits_{-1 \leqslant u \leqslant 0} |K(u)| \displaystyle\int_x^{x+1} dg(y).$

SPECIAL TAUBERIAN THEOREMS

From this we may readily conclude that for $-\infty < x < \infty$ there exists a K such that

(17·17) $$\int_x^{x+1} dg(y) < K.$$

For this we have only to remember that this integral is from its definition bounded for any finite range of x, and vanishes for $x < -1$. Combining these facts with (17·16), the result is clear.

If we remember that g is monotone increasing, we see that (10·05) is bounded. Hence in case there is no real u for which

(17·18) $$0 = \frac{1}{\sqrt{2\pi}} \int_{-\infty}^{\infty} K_1(x) e^{-iux} dx$$
$$= \frac{1}{\sqrt{2\pi}} \int_0^{\infty} \frac{d}{d\xi} \left(\frac{\xi e^{-\xi}}{e^{-\xi} - 1} \right) \xi^{iu} d\xi,$$

(10·07) follows for any $K_2(x)$ belonging to M_1. In particular, let

$$K_{21}(x) = 0 \quad (-\infty < x < -\epsilon); \quad K_{21}(x) = \frac{x+\epsilon}{\epsilon} \quad (-\epsilon \leqslant x < 0);$$
$$K_{21}(x) = e^{-x} \quad (0 \leqslant x);$$
$$K_{22}(x) = 0 \quad (-\infty < x < 0); \quad K_{22}(x) = \frac{x}{\epsilon} e^{-\epsilon} \quad (0 \leqslant x < \epsilon);$$
$$K_{22}(x) = e^{-x} \quad (\epsilon \leqslant x).$$

We have $$\int_{-\infty}^{\infty} K_{21}(x) dx = 1 + \epsilon/2;$$
$$\int_{-\infty}^{\infty} K_{22}(x) dx = e^{-\epsilon}(1 + \epsilon/2).$$

(10·15) thus becomes

$$1 + \epsilon/2 = \lim_{x \to \infty} \int_{-\infty}^{\infty} K_{21}(x-y) dg(y)$$
$$= \lim_{x \to \infty} \left\{ \int_{-\infty}^x e^{y-x} dg(y) + \int_x^{x+\epsilon} \frac{\epsilon - y + x}{\epsilon} dg(y) \right\}$$
$$\geqslant \varlimsup_{x \to \infty} \int_{-\infty}^x e^{y-x} dg(y)$$
$$= \varlimsup_{N \to \infty} \frac{1}{N} \int_0^N \eta \, dg(\log \eta)$$
$$= \varlimsup_{N \to \infty} \frac{1}{N} \sum_{n=1}^N \Lambda(n),$$

SPECIAL TAUBERIAN THEOREMS

and

$$e^{-\epsilon}(1+\epsilon/2) = \lim_{x\to\infty} \int_{-\infty}^{\infty} K_{22}(x-y)\,dg(y)$$

$$= \lim_{x\to\infty} \left\{ \int_{-\infty}^{x-\epsilon} e^{y-x}\,dg(y) + \int_{x-\epsilon}^{x} \frac{x-y}{\epsilon} e^{-\epsilon}\,dg(y) \right\}$$

$$= \lim_{x\to\infty} \left\{ \int_{-\infty}^{x} e^{y-x}\,dg(y) - \int_{x-\epsilon}^{x} \left(e^{y-x} - \frac{x-y}{\epsilon} e^{-\epsilon}\right) dg(y) \right\}$$

$$\leqslant \varlimsup_{x\to\infty} \int_{-\infty}^{x} e^{y-x}\,dg(y)$$

$$= \varlimsup_{N\to\infty} \frac{1}{N} \sum_{n=1}^{N} \Lambda(n).$$

Since ϵ is arbitrarily small and since, as $\epsilon \to 0$,

$$1 + \epsilon/2 \to 1, \quad e^{-\epsilon}(1+\epsilon/2) \to 1,$$

it follows that
$$1 \geqslant \varlimsup_{N\to\infty} \frac{1}{N} \sum_{n=1}^{\infty} \Lambda(n);$$

$$1 \leqslant \varliminf_{N\to\infty} \frac{1}{N} \sum_{n=1}^{\infty} \Lambda(n).$$

This is however impossible unless

(17·19) $$1 = \lim_{N\to\infty} \frac{1}{N} \sum_{n=1}^{N} \Lambda(n),$$

which we have shown to be equivalent to theorem 15.

§18. The Lambert-Tauber Theorem.

The only gap remaining in our proof of the prime-number theorem is the demonstration that the function in (17·18) has no real zeros. This function is

(18·01) $$\frac{1}{\sqrt{2\pi}} \int_0^\infty \frac{d}{d\xi}\left(\frac{\xi e^{-\xi}}{e^{-\xi}-1}\right) \xi^{iu}\,d\xi$$

$$= \lim_{\lambda\to 0} \frac{1}{\sqrt{2\pi}} \int_0^\infty \frac{d}{d\xi}\left(\frac{\xi e^{-\xi}}{e^{-\xi}-1}\right) \xi^{iu+\lambda}\,d\xi$$

$$= \lim_{\lambda\to 0} \frac{iu+\lambda}{\sqrt{2\pi}} \int_0^\infty \frac{\xi^{iu+\lambda} e^{-\xi}}{1-e^{-\xi}}\,d\xi$$

SPECIAL TAUBERIAN THEOREMS

$$= \lim_{\lambda \to 0} \frac{iu+\lambda}{\sqrt{2\pi}} \int_0^\infty \xi^{iu+\lambda} (e^{-\xi} + e^{-2\xi} + e^{-3\xi} + \ldots) d\xi$$

$$= \lim_{\lambda \to 0} \frac{iu+\lambda}{\sqrt{2\pi}} \left[\int_0^\infty \xi^{iu+\lambda} e^{-\xi} d\xi \right.$$

$$\left. + \frac{1}{2^{iu+\lambda+1}} \int_0^\infty \xi^{iu+\lambda} e^{-\xi} d\xi + \frac{1}{3^{iu+\lambda+1}} \int_0^\infty \xi^{iu+\lambda} e^{-\xi} d\xi + \ldots \right]$$

$$= \lim_{\lambda \to 0} \frac{iu+\lambda}{\sqrt{2\pi}} \Gamma(iu+\lambda+1) \sum_1^\infty n^{-iu-\lambda-1}$$

$$= \lim_{\lambda \to 0} \frac{iu+\lambda}{\sqrt{2\pi}} \Gamma(iu+\lambda+1) \zeta(iu+\lambda+1),$$

where $\zeta(u)$ is the Riemann zeta function, defined by

(18·02) $\qquad \zeta(w) = \sum_1^\infty n^{-w} \qquad (R(w) > 1).$

It will be noted that

(18·03) $\qquad \dfrac{1}{\sqrt{2\pi}} \int_0^\infty \dfrac{d}{d\xi} \left(\dfrac{\xi e^{-\xi}}{e^{-\xi} - 1} \right) \xi^w d\xi$

converges and defines an analytic function of w for $R(w) > 0$. We shall assume it to be known that $1/\Gamma(u)$ is an entire function, and that $\Gamma(u)$ has simple poles for $u = 0, -1, -2, \ldots$ and no other finite singularities. It will then follow from (18·01), (18·02), and (18·03) that $(w-1)\zeta(w)$ is equal to an analytic function which has no finite singularities for $R(w) > 0$, and that we may hence continue $\zeta(w)$ analytically over this region. It will be free from any singularity other than a pole of order 1 at $w = 1$. This pole will actually exist, for otherwise we should have

$$\infty > \lim_{w \to 1+0} \sum_1^\infty n^{-w} = \sum_1^\infty n^{-1},$$

which we know to be false. Thus by (18·01) the non-vanishing of (17·08) is equivalent to the non-vanishing of $\zeta(w)$ for $R(w) = 1$, and this is what we now have to prove to complete the proof of the Hadamard-de la Vallée Poussin prime-number theorem (theorem 15).

SPECIAL TAUBERIAN THEOREMS

Let us note that if $R(w) > 1$,

$$\zeta(w) = \sum_{1}^{\infty} n^{-w} = 1 + \Sigma p_1^{-k_1 w} p_2^{-k_2 w} \ldots p_\nu^{-k_\nu w}$$
$$= (1 + 2^{-w} + 2^{-2w} + \ldots + 2^{-k_1 w} + \ldots)$$
$$(1 + 3^{-w} + 3^{-2w} + \ldots)(1 + 5^{-w} + 5^{-2w} + \ldots) \ldots$$
$$= \frac{1}{(1 - 2^{-w})(1 - 3^{-w})(1 - 5^{-w}) \ldots (1 - p^{-w}) \ldots}$$
$$= \prod_p (1 - p^{-w})^{-1},$$

where the product is taken over all primes. Thus

(18·04) $\qquad \log \zeta(w) = - \sum_p \log(1 - p^{-w})$
$$= \sum_p \left(p^{-w} + \frac{p^{-2w}}{2} + \frac{p^{-3w}}{3} + \ldots \right)$$
$$= \sum_{1}^{\infty} \frac{\Lambda(n)}{\log n} n^{-w}.$$

We now wish to establish

Lemma 15$_2$. *Let $\gamma(x)$ be a monotone increasing function, and let*

$$\int_{1+0}^{\infty} x^{-w} d\gamma(x) = \phi(w)$$

converge for $R(w) > 1$. Let

$$F(w) = e^{\phi(w)} (w - 1)^A \qquad [0 < A < 2^{\frac{1}{2}}]$$

when continued analytically, be regular for $R(w) = 1$, and let it not vanish for $w = 1$. Then $F(w)$ will have no zeros for $R(w) = 1$.

It will follow from this lemma—which is in essence due to Hadamard[*], that if we put

$$\gamma(x) = \sum_{1}^{[x]} \frac{\Lambda(n)}{\log n};$$
$$\phi(w) = \log \zeta(w);$$
$$F(w) = \zeta(w)(w - 1);$$
$$A = 1,$$

then $\zeta(w)$ cannot vanish for $R(w) = 1$, and thus the proof of theorem 15 will be complete.

---
[*] Landau 2, p. 166.

SPECIAL TAUBERIAN THEOREMS

Now, as to the proof: let us notice that

(18·05) $\quad 0 \leqslant (1+\sqrt{2}\cos\phi)^2 = 1 + 2\sqrt{2}\cos\phi + 2\cos^2\phi$
$$= 2 + 2\sqrt{2}\cos\phi + \cos 2\phi.$$

Again, from the regularity of $F(w)$, it follows that

(18·06) $\qquad\qquad \phi(w) + A\log(w-1)$

is regular for $R(w) = 1$, except for logarithmic singularities where the real part of (18·06) grows *negatively* infinite. These we shall call negative logarithmic singularities. The point $w = 1$ is a regular point of (18·06).

Now, $\quad R(\phi(w)) = \int_{1+0}^{\infty} x^{-R(w)} \cos(I(w)\log x)\, d\gamma(x),$

so that, if v is real,

$$R(\phi(1+\epsilon+iv)) = \int_{1+0}^{\infty} x^{-(1+\epsilon)} \cos(v\log x)\, d\gamma(x);$$

$$R(\phi(1+\epsilon+2iv)) = \int_{1+0}^{\infty} x^{-(1+\epsilon)} \cos(2v\log x)\, d\gamma(x);$$

and $\qquad\qquad \phi(1+\epsilon) = \int_{1+0}^{\infty} x^{-(1+\epsilon)}\, d\gamma(x).$

Thus, by (18·05),

$2\phi(1+\epsilon) + 2\sqrt{2}R(\phi(1+\epsilon+iv)) + R(\phi(1+\epsilon+2iv))$

$= \int_{1+0}^{\infty} x^{-(1+\epsilon)} (2 + 2\sqrt{2}\cos(v\log x) + \cos(2v\log x))\, d\gamma(x) \geqslant 0.$

Hence

$R(\phi(1+\epsilon+iv)) \geqslant -2^{-\frac{1}{2}}\phi(1+\epsilon) - 2^{-\frac{3}{2}}R\phi(1+\epsilon+2iv),$

and, if $\epsilon < 1$,

$$\frac{R(\phi(1+\epsilon+iv))}{\log\epsilon} \leqslant -2^{-\frac{1}{2}}\frac{\phi(1+\epsilon)}{\log\epsilon} - 2^{-\frac{3}{2}}\frac{R\phi(1+\epsilon+2iv)}{\log\epsilon}.$$

As ϵ tends to 0 from above, $R\phi(1+\epsilon+2iv)/\log\epsilon$ remains positive, while
$$\phi(1+\epsilon)/\log\epsilon \to -A.$$
Thus

(18·065) $\qquad \overline{\lim_{\epsilon \to +0}} \dfrac{R(\phi(1+\epsilon+iv))}{\log\epsilon} \leqslant 2^{-\frac{1}{2}} A < 1.$

SPECIAL TAUBERIAN THEOREMS

On the other hand, $F(w)$ is analytic for $w = 1 + iv$. This means that there is a positive or zero integer n such that

$$(18\text{·}07) \qquad \lim_{\epsilon \to +0} \frac{F(1 + \epsilon + iv)}{\epsilon^n} = B,$$

where B is finite and does not vanish. Hence, if $v \neq 0$,

$$\log B = \lim_{\epsilon \to 0} (\log F(1 + \epsilon + iv) - n \log \epsilon)$$
$$= \lim_{\epsilon \to 0} (\phi(1 + \epsilon + iv) + A \log(\epsilon + iv) - n \log \epsilon)$$
$$= A \log iv + \lim_{\epsilon \to 0} (\phi(1 + \epsilon + iv) - n \log \epsilon).$$

If we divide by $\log \epsilon$ and proceed to the limit, this gives us

$$0 = \lim_{\epsilon \to 0} \left(\frac{\phi(1 + \epsilon + iv) - n \log \epsilon}{\log \epsilon} \right),$$

or $\quad \lim\limits_{\epsilon \to +0} \dfrac{R(\phi(1 + \epsilon + iv))}{\log \epsilon} = R \lim\limits_{\epsilon \to 0} \dfrac{\phi(1 + \epsilon + iv)}{\log \epsilon} = n.$

If we combine this with (18·065), and remember that n is an integer, we see that it can only be 0, and by (18·07),

$$\lim_{\epsilon \to +0} F(1 + \epsilon + iv) = B.$$

Thus $1 + iv$ is not a zero of $F(w)$, and hence $F(w)$ can have no zeros for $R(w) = 1$. This establishes lemma 15_2, and completes the proof of theorem 15.

The history of the application of Lambert series to the study of the distribution of primes is rather interesting. The particular series most studied by Lambert, and known as *the* Lambert series, is

$$(18\text{·}08) \qquad \sum_1^\infty \frac{x^n}{1 - x^n} = x + 2x^2 + 2x^3 + 3x^4 + 2x^5 + 4x^6 + \ldots,$$

where the coefficient of x^m in the equivalent power series represents the number of distinct factors of m, including 1 and m itself. This is 2 when and only when m is a prime, and so much hope was at one time placed in this series as a due to the distribution of the primes. This hope has been characterized by Knopp* as "recht verführerisch"—thoroughly misleading—and

---

* Knopp 1.

indeed (18·08) itself has proved sterile of results in prime-number theory. This has also been the case until recently with Lambert series of more general form. In 1921 Hardy and Littlewood* proved the first positive result concerning the relation between the theory of the distribution of the primes and the theory of Lambert series. They established that the Hadamard-de la Vallée Poussin theorem (theorem 15) was deducible from the following Tauberian theorem concerning Lambert series; *if* $na_n > -K$, *and*

$$\lim_{x \to 1-0} \sum_{0}^{\infty} \frac{na_n x^n (1-x)}{1-x^n} = A,$$

then $\Sigma a_n = A$. This theorem however they did not establish autonomously, although it was established as a corollary of a theorem proved by Landau by methods somewhat more difficult than those required for the Hadamard-de la Vallée Poussin theorem.

The method of this chapter and of previous papers by the author† may also be viewed as the reduction of a Tauberian theorem on Lambert series to a known theorem in prime-number theory—in this case, to lemma 15_2, which allows us to show that the function $\zeta(w)$ has no zeros on the line $R(w) = 1$. This has always been recognized as the crucial step in the proof of theorem 15, although generally it has been associated with some very mild proposition as to the behaviour of the Riemann zeta function at infinity. This we shall see in the next section. However, there are extant proofs of theorem 15 in which the sole non-elementary property of the Riemann zeta function which comes into play is its non-vanishing on the 1-line‡, and to that extent we may say that the logic of the proof given here does not really go much further than that of the Hardy-Littlewood paper. The one claim which may legitimately be made for the methods of the present chapter is that they furnish an independent method of reducing theorem 15 to a lemma of the type of 15_2.

* Hardy and Littlewood 1.
† Wiener 2, 4.
‡ Cf. the discussion of this point in A. E. Ingham's forthcoming Cambridge Tract on the Distribution of Prime Numbers.

SPECIAL TAUBERIAN THEOREMS 125

§ 19. Ikehara's Theorem.

One of the most straightforward methods of proving the prime-number theorem depends on a lemma due to Landau*. This lemma reads as follows:

Landau's Lemma. *Let*

(19·01) $\quad F(x) = \sum\limits_{n=1}^{\infty} a_n n^{-x}, \quad [R(x) > 1],$

and let

(19·02) $\quad a_n \geqslant 0, \quad [n = 1, 2, \ldots].$

Let $F(x)$ be analytically continuable on to $R(x) = 1$, and let it there be free from singularities, except for a pole of order one at $x = 1$, with principal part $A/(x-1)$. Let there be some finite α for which

(19·03) $\quad F(x) = O(|x|^\alpha)$

in the right half-plane $R(x) \geqslant 1$. Then

(19·04) $\quad A = \lim\limits_{n \to \infty} \frac{1}{n} \sum\limits_{1}^{n} a_k.$

To apply this to the demonstration of the prime-number theorem, we take (18·04) and differentiate it term by term, as is legitimate for $R(w) > 1$, for a uniformly convergent series of analytic functions may be differentiated term by term at an interior point of the region of uniform convergence. If we change the sign, we get

$$-\zeta'(w)/\zeta(w) = \sum\limits_{1}^{\infty} \Lambda(n) n^{-w}.$$

Furthermore, we have seen that a constant B exists (it is actually 1) such that

$$\zeta(w) - B/(w-1)$$

is analytic at $w = 1$. The same is consequently true of

$$\zeta'(w) + B/(w-1)^2.$$

From this it immediately results that

$$-\zeta'(w)/\zeta(w) - 1/(w-1)$$

is analytic at $w = 1$, so that $-\zeta'(w)/\zeta(w)$ has a pole of order 1 at $w = 1$, with principal part $1/(w-1)$.

* Landau 3.

126 SPECIAL TAUBERIAN THEOREMS

We shall not set forth the proof of the fact that

$$F(x) = -\zeta'(x)/\zeta(x)$$

satisfies (19·03), inasmuch as we shall show that this condition may be dispensed with. Actually much more is known, and it has been shown* that if $R(s) \geq 1$ and $|s|$ is large,

$$\left|\frac{\zeta'(s)}{\zeta(s)}\right| < c \log^9 t.$$

The proof that $F(x)$ satisfies a condition as lenient as (19·03) offers very little difficulty. Condition (19·02) is of course fulfilled.

Assuming then that Landau's lemma has been established, and that $\zeta'(x)/\zeta(x)$ has been demonstrated to satisfy the hypothesis, we see that (19·04) becomes

(17·085) $$1 = \lim_{N \to \infty} \frac{1}{N} \sum_{n=1}^{N} \Lambda(n).$$

We have however already established the equivalence of this with the Hadamard-de la Vallée Poussin theorem. Thus we have an alternative demonstration of theorem 15.

It will be seen that condition (19·03) is a very weak restriction on $F(x)$, and is not apparently germane to the subject at all. Landau's lemma cannot be regarded as in a satisfactory state until either (19·03) is shown to be the weakest restriction on the order of $F(x)$ that is sufficient to guarantee the truth of the lemma, or else the true restriction is found. In this sense, the following theorem of Hardy and Littlewood† may be regarded as an advance on that of Landau.

The Hardy-Littlewood Theorem. *Let λ_n be an increasing real sequence. Let* (i) *the series $\Sigma a_n \lambda_n^{-s}$ be absolutely convergent for $R(s) > \sigma_0 > 0$;* (ii) *the function $F(s)$ defined by the series be regular for $R(s) > c$ where $0 < c \leq \sigma_0$ and continuous for $R(s) \geq c$, except for a simple pole with residue g at $s = c$;*

(19·05) (iii) $F(s) = O(e^{C|t|})$

---
* Landau 2, vol. 1, p. 179.
† Hardy and Littlewood 2.

SPECIAL TAUBERIAN THEOREMS

for some finite C, uniformly for $\sigma \geqslant c$;

(iv) $\quad \lambda_n/\lambda_{n-1} \to 1$;

(v) a_n *be real, and satisfy one of the inequalities*

$$a_n > -K\lambda_n^{c-1}(\lambda_n - \lambda_{n-1}); \quad a_n < K\lambda_n^{c-1}(\lambda_n - \lambda_{n-1}),$$

or complex, and of the form

$$O\{\lambda_n^{c-1}(\lambda_n - \lambda_{n-1})\}.$$

Then

(19·06) $\quad A_n = a_1 + a_2 + \ldots + a_n \sim g\lambda_n^c/c.$

Here we take

$$s = \sigma + it.$$

In the case where $c = \sigma_0 = 1$, $\lambda_n = n$, $K = 0$, $g = A$ the lemma reads:

Let (19·01) and (19·02) hold. Let $F(x)$ be regular on $R(x) = 1$, except for a pole of order 1 at $x = 1$, with principal part $A/(x-1)$. Let (19·05) hold for some finite C, uniformly for $\sigma \geqslant 1$. Then (19·04) holds.

It will be seen that (19·03) is replaced by the weaker condition (19·05). The real truth however is that the proposition last stated is true without even this weakened hypothesis concerning the order of magnitude of $F(s)$, and that any such hypothesis may merely be cancelled. The true theorem is that of Ikehara*, and reads as follows when stated in its Stieltjes' form:

THEOREM 16. *Let $\alpha(x)$ be a monotone increasing function, and let*

$$\int_{1+0}^{\infty} x^{-u} d\alpha(x) = f(u)$$

converge for $R(u) > 1$. Let

(19·065) $\qquad f(u) - \dfrac{A}{u-1} = g(u)$

converge over any finite interval of the line $R(u) = 1$ uniformly to a finite limit as $R(u) \to 1 + 0$. Then

$$\alpha(N) \sim NA$$

as $N \to \infty$.

* Ikehara 1.

128 SPECIAL TAUBERIAN THEOREMS

To prove this, let us put
$$\beta(\xi) = \alpha(e^\xi)e^{-\xi} + \int_0^\xi e^{-\xi}\alpha(e^\xi)\,d\xi - A\xi$$
for $\xi > 0$. Thus

(19·07) $\qquad d\beta(\xi) = e^{-\xi}d\alpha(e^\xi) - A\,d\xi.$

Let us assume—what is no essential restriction—that

(19·08) $\qquad \beta(x) = \beta(+0) \qquad (-\infty \leqslant x \leqslant 0).$

Then the $g(u)$ of (19·065) becomes
$$g(u) = \int_{-\infty}^{\infty} e^{(1-u)\xi}\,d\beta(\xi), \qquad [R(u) > 1].$$

What we wish to prove may be written

(19·085) $\qquad \lim_{\eta \to \infty} \int_{-\infty}^{\eta} e^{\xi-\eta}\,d\beta(\xi) = 0.$

If $\epsilon > 0$ and η is real, we have

(19·09) $\displaystyle\int_{-B}^{B} \left(1 - \frac{|u|}{B}\right) g(iu + \epsilon + 1)\, e^{iu\eta}\,du$

$\qquad = \displaystyle\int_{-B}^{B}\left(1 - \frac{|u|}{B}\right) du \int_{-\infty}^{\infty} e^{iu(\eta-\xi)}\, e^{-\epsilon\xi}\,d\beta(\xi).$

As this double integral is absolutely convergent, it becomes

(19·10) $\displaystyle\int_{-\infty}^{\infty} e^{-\epsilon\xi}\,d\beta(\xi) \int_{-B}^{B}\left(1 - \frac{|u|}{B}\right) e^{iu(\eta-\xi)}\,du$

$\qquad = \displaystyle\int_{-\infty}^{\infty} \frac{2(1 - \cos B(\eta - \xi))}{B(\eta - \xi)^2}\, e^{-\epsilon\xi}\,d\beta(\xi).$

Now,

(19·11) $\displaystyle\lim_{\epsilon \to 0}\int_{-B}^{B}\left(1 - \frac{|u|}{B}\right) g(iu + \epsilon + 1)\, e^{iu\eta}\,du$

$\qquad = \displaystyle\int_{-B}^{B}\left(1 - \frac{|u|}{B}\right) g(iu + 1)\, e^{iu\eta}\,du.$

Thus $\displaystyle\lim_{\epsilon \to 0}\int_{-\infty}^{\infty} \frac{2(1 - \cos B(\eta - \xi))}{B(\eta - \xi)^2}\, e^{-\epsilon\xi}\,d\beta(\xi)$

exists. Furthermore,

(19·12) $\displaystyle\lim_{\epsilon \to 0}\int_0^\infty \frac{2(1 - \cos B(\eta - \xi))}{B(\eta - \xi)^2}\, e^{-\epsilon\xi} A\,d\xi$

$\qquad = \displaystyle\int_0^\infty \frac{2(1 - \cos B(\eta - \xi))}{B(\eta - \xi)^2}\, A\,d\xi$

SPECIAL TAUBERIAN THEOREMS 129

by bounded convergence. Hence, by (19·07),

$$\lim_{\epsilon \to 0} \int_0^\infty \frac{2(1 - \cos B(\eta - \xi))}{B(\eta - \xi)^2} e^{-(\epsilon+1)\xi} d\alpha(e^\xi)$$

exists. If we remember that $e^{-\xi} d\alpha(\xi)$ is non-negative, that $e^{-\epsilon\xi}$ is monotone increasing to 1 for each positive ξ as $\epsilon \to 0$, and that

$$\frac{2(1 - \cos B(\eta - \xi))}{B(\eta - \xi)^2} \geq 0,$$

we may apply X_{39}, and obtain

(19·13) $\quad \displaystyle\int_0^\infty \frac{2(1 - \cos B(\eta - \xi))}{B(\eta - \xi)^2} e^{-\xi} d\alpha(e^\xi)$

$$= \lim_{\epsilon \to 0} \int_0^\infty \frac{2(1 - \cos B(\eta - \xi))}{B(\eta - \xi)^2} e^{-(\epsilon+1)\xi} d\alpha(e^\xi).$$

Combining (19·12) and (19·13), we obtain

$$\lim_{\epsilon \to 0} \int_{-\infty}^\infty \frac{2(1 - \cos B(\eta - \xi))}{B(\eta - \xi)^2} e^{-\epsilon\xi} d\beta(\xi)$$

$$= \int_{-\infty}^\infty \frac{2(1 - \cos B(\eta - \xi))}{B(\eta - \xi)^2} d\beta(\xi),$$

in view of (19·07) and (19·08). Thus, by (19·09), (19·10), and (19·11),

(19·14) $\quad \displaystyle\int_{-B}^B \left(1 - \frac{|u|}{B}\right) g(iu + 1) e^{iu\eta} du$

$$= \int_{-\infty}^\infty \frac{2(1 - \cos B(\eta - \xi))}{B(\eta - \xi)^2} d\beta(\xi).$$

Thus by (19·07)

$$\int_{-\infty}^\infty \frac{2(1 - \cos B(\eta - \xi))}{B(\eta - \xi)^2} e^{-\xi} d\alpha(\xi)$$

$$= 2\pi A + \int_{-B}^B \left(1 - \frac{|u|}{B}\right) g(iu + 1) e^{iu\eta} du.$$

As $e^{-\xi} d\alpha(\xi)$ is non-negative, if $B < \pi$,

$$\int_\eta^{\eta+1} e^{-\xi} d\alpha(e^\xi) < \frac{\pi^2}{4B} \left\{ 2\pi A + \int_{-B}^B \left(1 - \frac{|u|}{B}\right) |g(iu+1)| du \right\}$$

$$< \infty,$$

and since α is monotone, it again follows from (19·07) that

$$\int_\eta^{\eta+1} |d\beta(\xi)| < A + \frac{\pi^3 A}{2B} + \frac{\pi^2}{4B}\int_{-B}^{B}\left(1-\frac{|u|}{B}\right)|g(iu+1)|du < \infty.$$

By the Riemann-Lebesgue theorem and (19·14),

$$\lim_{\eta\to\infty}\int_{-\infty}^{\infty}\frac{2(1-\cos B(\eta-\xi))}{B(\eta-\xi)^2}d\beta(\xi)=0.$$

Moreover, the Fourier transform of

$$\frac{2(1-\cos B\xi)}{B\xi^2}$$

is

$$\frac{1}{\sqrt{2\pi}}\int_{-\infty}^{\infty}\frac{2(1-\cos B\xi)}{B\xi^2}e^{-iu\xi}d\xi = \sqrt{2\pi}\left(1-\frac{|u|}{B}\right) \quad \text{if} \quad |u|<B;$$

$$= 0 \quad \text{if} \quad |u|>B.$$

There is no value of u for which this vanishes for all B. We may thus apply theorem 7, which yields the result that if $K_2(\xi)$ belongs to M_1, then

(19·15) $$\lim_{\eta\to\infty}\int_{-\infty}^{\infty}K_2(\eta-\xi)d\beta(\xi)=0.$$

Proposition (19·085) is the particular case of this for which

(19·16) $\quad K_2(\xi)=e^{-\xi} \ (\xi>0); \quad K_2(\xi)=0 \ (\xi<0).$

This was however the result to be proved. We have thus established theorem 16.

The generalized form of the Landau theorem corresponding to theorem 16 is obtained by taking for $\alpha(x)$ a step-function constant between integral arguments. It reads:

THEOREM 17. *Let* (19·01) *and* (19·02) *hold. Let* $F(x)$ *be analytically continuable on to* $R(x)=1$, *and let it there be free of singularities, except for a pole of order one at* $x=1$, *with principal part* $A/(x-1)$. *Then* (19·04) *holds.*

In generalizing the Hardy-Littlewood theorem I have been unable to eliminate (iii) without strengthening (iv). If we replace (iv) by (iv'), $\lambda_n-\lambda_{n-1}$ is bounded; we have

THEOREM 18. *In the Hardy-Littlewood theorem, conditions* (i), (ii), (iv'), *and* (v), *without* (iii), *are sufficient for* (19·06).

SPECIAL TAUBERIAN THEOREMS 131

A less stringent substitute for (iv) is

(iv'') $$\Sigma\,(\lambda_n-\lambda_{n-1})^2\,\lambda_{n-1}^{-2}<\infty.$$

This theorem is not so obvious as the corresponding Landau theorem. We may suppose without real restriction that all λ_n exceed 1. Let us put

$$\alpha(x)=\sum_{\lambda_n\leqslant x}a_n\lambda_n^{1-c};$$
(19·165) $$u=s+1-c;$$
$$\sigma_1=\sigma_0+1-c.$$

Then (i) asserts that

(19·17) $$\int_{1+0}^{\infty}x^{-u}d\alpha(x)=f(u)=F(u+c-1)$$

converges absolutely for $R(u)>\sigma_1$. Proposition (ii) asserts that $f(u)$, when continued analytically, is free from singularities for $R(u)>1$, and is continuous for $R(u)=1$, except for a simple pole at $u=1$ with residue g. Proposition (v) implies that either $\alpha(x)$ is real, and satisfies one of the conditions

(19·18) $$\int_n^{n+1}|d\alpha(x)|-\alpha(n+1)+\alpha(n)<2K$$

or

(19·19) $$\int_n^{n+1}|d\alpha(x)|+\alpha(n+1)-\alpha(n)<2K,$$

or complex, and satisfies the condition

(19·20) $$\int_n^{n+1}|d\alpha(x)|<K$$

for some constant K. In case (19·20), if $R(u)>1+\epsilon$,

$$\int_n^{n+1}|x^{-u}d\alpha(x)|<Kn^{-1-\epsilon},$$

and (19·17) converges absolutely for $R(u)>1$, as we see by comparing it with the series

$$\sum_1^{\infty}\int_n^{n+1}|x^{-u}d\alpha(x)|<K\sum_1^{\infty}n^{-1-\epsilon}.$$

In case (19·18) we may write

$$\alpha(x)=\alpha_1(x)+\alpha_2(x),$$

SPECIAL TAUBERIAN THEOREMS

where
$$\alpha_1(x) = \tfrac{1}{2}\left(\alpha(x) + \int_{1+0}^{x} |d\alpha(\xi)|\right);$$
$$\alpha_2(x) = \tfrac{1}{2}\left(\alpha(x) - \int_{1+0}^{x} |d\alpha(\xi)|\right).$$

Then $\alpha_1(x)$ will be monotone increasing, while

(19·201) $\quad \int_n^{n+1} |d\alpha_2(x)| = \tfrac{1}{2} \int_n^{n+1} \left| d\left\{ \int_{1+0}^{x} |d\alpha(\xi)| - \alpha(x) \right\} \right|.$

Since the expression under the outer d on the right side of (19·201) is monotone increasing, this gives us

$$\int_n^{n+1} |d\alpha_2(x)| = \tfrac{1}{2} \int_n^{n+1} d\left\{ \int_{1+0}^{x} |d\alpha(\xi)| - \alpha(x) \right\}$$
$$= \tfrac{1}{2} \left\{ \int_n^{n+1} |d\alpha(x)| - \alpha(n+1) + \alpha(n) \right\}$$
$$< K.$$

Thus, as in case (19·20) with $\alpha(x)$,

(19·202) $\quad \int_{1+0}^{\infty} x^{-u} d\alpha_2(x)$

converges absolutely for $R(u) > 1$. As $\alpha_1(x)$ is increasing,

(19·203) $\quad \int_{1+0}^{\infty} x^{-u} d\alpha_1(x)$

is a decreasing function of u. We know that (19·203) converges for $R(u) > \sigma_1$. Let σ_1 be the least value of $R(u)$ for which this is true. It follows that

$$\lim_{u \to \sigma_1 + 0} \int_{1+0}^{\infty} x^{-u} d\alpha_1(x) = \infty.$$

For, if u is real,

(19·21) $\quad \sum_{n=0}^{\infty} 2^{-nu} \int_{2^n}^{2^{n+1}} d\alpha_1(x)$

converges or diverges with (19·20), and bears to it a ratio somewhere between 1 and 2^u, inclusive. However, (19·21) is a power-series in 2^{-u} with non-negative coefficients, and such a series must converge up to a real infinite singularity of the function it represents[*]. Thus the function represented by (19·20) must have an infinite singularity on its ordinate at the abscissa of

---

[*] Landau, *Neuere Ergebnisse der Funktionentheorie* (2nd edition).

SPECIAL TAUBERIAN THEOREMS 133

convergence, and, *a fortiori*, must have a singularity at the real point with this abscissa.

The left-most singularity on the real axis is however $\leqslant 1$. Thus, the abscissa of convergence is $\leqslant 1$, and $\sigma_1 = 1$. It is not here necessary to distinguish between $\alpha(x)$ and $\alpha_1(x)$, in view of the absolute convergence of (19·202) for $R(u) > 1$.

Now we have established in cases (19·18) and (19·20) that (19·17) converges absolutely for $R(u) > 1$. Case (19·19) is quite like (19·18). From now on, we shall devote our attention to case (19·18). Let us put*

$$\alpha_3(x) = \sum_{\lambda_n \leqslant x} \{a_n \lambda_n^{1-c} + K(\lambda_n - \lambda_{n-1})\}.$$

This is a step-function whose jumps occur at $x = \lambda_n$ and are of magnitude $a_n \lambda_n^{1-c} + K(\lambda_n - \lambda_{n-1})$. We wish to show that

$$(19·22) \qquad \int_{1+0}^{\infty} x^{-u} d\alpha_3(x)$$

also converges absolutely for $R(u) > 1$. This amounts to showing that

$$\Sigma \lambda_n^{-u} \{a_n \lambda_n^{1-c} + K(\lambda_n - \lambda_{n-1})\}$$

converges for $R(u) > 1$, and by (19·165) and condition (i), to showing that

$$(19·23) \qquad \Sigma \lambda_n^{-u} (\lambda_n - \lambda_{n-1}) = \psi(u)$$

converges for $R(u) > 1$. That this is the case, however, results from the fact that (19·23) is dominated by the integral

$$\int_1^{\infty} \lambda^{-u} d\lambda = 1/(u-1)$$

which converges for $R(u) > 1$.

We have

(19·24)

$$\psi(u) - 1/(u-1) = e(u) + \sum_2^{\infty} \left\{ (\lambda_n - \lambda_{n-1}) \lambda_n^{-u} - \int_{\lambda_{n-1}}^{\lambda_n} \lambda^{-u} du \right\},$$

where $e(u)$ is an entire function. The expression under the Σ sign does not exceed in modulus

$$(\lambda_n - \lambda_{n-1})(\lambda_{n-1}^{-u} - \lambda_n^{-u}) \leqslant u(\lambda_n - \lambda_{n-1})^2 \lambda_{n-1}^{-u-1},$$

* Here we take λ_0 to be 0.

by the theorem of the mean. In view of (iv′), this is dominated by

$$\text{const. } u \, (\lambda_n - \lambda_{n-1}) \lambda_{n-1}^{-u-1},$$

and (19·24) converges to an analytic function for $R(u) > 0$. Thus (19·22) converges for $R(u) > 1$, to a function which, when continued analytically, is continuous on $R(u) = 1$, except at $u = 1$ itself. At this point

$$\int_{1+0}^{\infty} x^{-u} d\alpha_3(x) - K\psi(u)$$

has a simple pole with residue g, and by (19·24), (19·22) has a simple pole with residue $g + K$. The function $\alpha_3(x)$ is monotone, and we are in a position to apply the methods of theorem 16, substituting $\alpha_3(x)$ for $\alpha(x)$.

We showed in the discussion of theorem 16 that the hypothesis is a sufficient condition for (19·15), whenever K_2 belongs to M_1. Instead of the K_2 of (19·16), let us put

$$K_2(\xi) = e^{-c\xi} \ (\xi > 0); \quad K_2(\xi) = 0 \ (\xi < 0).$$

The appropriate $\beta(\xi)$ will be

$$\alpha_3(e^{\xi}) e^{-\xi} + \int_0^{\xi} e^{-\xi} \alpha_3(e^{\xi}) d\xi - (g + K)\xi$$

for $\xi > 0$, and we shall have $\beta(\xi) = \beta(+0)$ for $\xi < 0$. We have established that

(19·25)
$$0 = \lim_{\eta \to \infty} \int_{-\infty}^{\infty} K_2(\eta - \xi) d\beta(\xi)$$

$$= \lim_{\eta \to \infty} \int_{+0}^{\eta} e^{c(\xi - \eta)} \{ e^{-\xi} d\alpha_3(e^{\xi}) - (g + K) d\xi \}$$

$$= \lim_{\nu \to \infty} \nu^{-c} \left\{ \int_{1+0}^{\nu} x^{c-1} d\alpha_3(x) - \frac{(g+K)(\nu^c - 1)}{c} \right\}$$

$$= \lim_{\nu \to \infty} \left\{ \nu^{-c} \sum_{\lambda_n \leqslant \nu} [a_n \lambda_n^{1-c} + K(\lambda_n - \lambda_{n-1})] \lambda_n^{c-1} - \frac{g+K}{c}(1 - \nu^{-c}) \right\}$$

$$= \lim_{N \to \infty} \left\{ \lambda_N^{-c} \sum_1^N a_n + K \lambda_N^{-c} \sum_1^N (\lambda_n - \lambda_{n-1}) \lambda_n^{c-1} - \frac{g+K}{c}(1 - \lambda_N^{-c}) \right\}.$$

SPECIAL TAUBERIAN THEOREMS

However, this is true for *any* sufficiently large K. Thus by combining two equations such as (19·25) linearly for different values of K, we see that we have separately

$$\lim_{N\to\infty}\left\{\lambda_N^{-c}\sum_1^N(\lambda_n-\lambda_{n-1})\lambda_n^{c-1}-\frac{1}{c}(1-\lambda_N^{-c})\right\}=0$$

and
$$\lim_{N\to\infty}\left\{\lambda_N^{-c}\sum_1^N a_n-\frac{g}{c}(1-\lambda_N^{-c})\right\}=0.$$

It follows that, since $g \neq 0$, $c \neq 0$,

$$\lambda_N^{-c}\sum_1^N a_n \sim \frac{g}{c}(1-\lambda_N^{-c}) \sim \frac{g}{c},$$

and hence that
$$\sum_1^N a_n \sim \frac{g}{c}\lambda_N^{-c},$$

which is identical with (19·06). Thus we have established theorem 18 in the case (19·18). The case (19·19) follows immediately by replacing a_n by $-a_n$, and the case (19·20) may be reduced to either of the above cases by a separate consideration of real and imaginary parts.

If we combine lemma 15_2, and theorem 16, we obtain the following interesting theorem:

THEOREM 19. *Let $\gamma(x)$ be a monotone increasing function, and let*

$$\int_{1+0}^{\infty} x^{-w} d\gamma(x) = \phi(w)$$

converge for $R(w) > 1$. Let

$$F(w) = e^{\phi(w)}(w-1)^A, \qquad [0 < A < 2^{\frac{1}{2}}],$$

when continued analytically, be regular for $R(w) = 1$, and let it not vanish for $w = 1$. Then

$$A = \lim_{N\to\infty}\frac{1}{N}\int_{1+0}^{N}\log x\,d\gamma(x).$$

For let us put $\int_{1+0}^{x}\log\xi\,d\gamma(\xi) = \alpha(x)$;

$$-\phi'(u) = f(u).$$

Then, if $R(u) > 1$,

$$\int_{1+0}^{\infty} x^{-u}\,d\alpha(x) = \int_{1+0}^{\infty} x^{-u}\log x\,d\gamma(x)$$

$$= -\frac{d}{du}\int_{1+0}^{\infty} x^{-u}\,d\gamma(x) = -\phi'(u) = f(u).$$

Again, $\log F(u) = \phi(u) + A \log(u-1)$,
so that $-F'(u)/F(u) = f(u) - A/(u-1)$.

This function is analytic for $R(u) = 1$, by lemma 15_2, so that it converges uniformly to a finite limit over any finite interval of the line $R(u) = 1$ as $R(u) \to 1$ from below. Thus the hypothesis of theorem 16 is satisfied, and

$$A = \lim_{N \to \infty} \frac{\alpha(N)}{N} = \lim_{N \to \infty} \frac{1}{N} \int_{1+0}^{N} \log x \, d\gamma(x),$$

which is the desired result.

If we put $\gamma(x) = \varpi(x)$,
then $e^{\phi(u)} = \zeta(u)$,

and the hypothesis of theorem 17 is manifestly satisfied. The conclusion then becomes

$$1 = \lim_{N \to \infty} \frac{1}{N} \int_{1+0}^{N} \log x \, d\varpi(x),$$

which we have already shown to be equivalent to the prime-number theorem.

We have thus shown in two distinct ways that the prime-number theorem, in the form due to Hadamard and de la Vallée Poussin, may be reduced to Tauberian form and thus established. This might lead to an expectation that the more refined theorems on the distribution of the primes may also be reduced to Tauberian form—in particular, those theorems in which the Riemann hypothesis is assumed. The author does not expect much from Tauberian methods in this wider field. It will be noticed in theorems 4 and 5 that in order to prove that a certain expression approaches a limit, we assume hypotheses which are trivially sufficient to secure its boundedness. That is, the whole refinement of the Tauberian method serves the purpose of changing an O into an o. In the case of theorem 15, that is all that is needed, for it may be proved by elementary—though not trivial—methods[*] that

$$\pi(n) - \frac{n}{\log n} = O\left(\frac{n}{\log n}\right),$$

[*] Landau 2, vol. 1, pp. 71-97.

SPECIAL TAUBERIAN THEOREMS

or, what is the same, that

$$\pi(n) = O\left(\frac{n}{\log n}\right),$$

while theorem 15 asserts that

$$\pi(n) - \frac{n}{\log n} = o\left(\frac{n}{\log n}\right).$$

In the case of the more refined theorems involving the Riemann hypothesis, such as

(19·26) $$\pi(n) - \frac{n}{\log n} = o(n^\alpha), \qquad (\tfrac{1}{2} < \alpha < 1),$$

the O proposition will be practically as hard to establish as the o proposition. In the case of (19·26) it will be

$$\pi(n) - \frac{n}{\log n} = O(n^\alpha),$$

and to establish this is to establish more than

$$\pi(n) - \frac{n}{\log n} = o(n^\beta), \qquad (\alpha < \beta < 1).$$

To establish this proposition for every α properly between $\tfrac{1}{2}$ and 1, and for every β properly between α and 1, is identically the same thing as establishing (19·26) for every α properly between $\tfrac{1}{2}$ and 1, and a Tauberian theorem here serves no purpose whatever.

Another way of looking at the matter is that a Tauberian theorem depends on the Fourier transform theory, which is a chapter from the theory of the functions of a real variable. We are thus confined to operations along a single line,—in the case of prime-number theory along a single ordinate in the plane of the Riemann zeta function. The more powerful methods of complex function theory allow us on the other hand to work in a whole strip of the complex plane—the critical strip of the Riemann zeta function. We may therefore proceed in one leap across a band of ordinates, and arrive at such a proposition as (19·26).

§ 20. The Mean Square Modulus of a Function.

THEOREM 20. *Let $f(x)$ be a measurable function for which*

(20·01) $$\frac{1}{2T}\int_{-T}^{T}|f(x)|^2\,dx$$

is bounded in T. Then

(20·02) $$\int_{-\infty}^{\infty}\frac{|f(x)|^2}{1+x^2}\,dx<\infty.$$

For if we integrate by parts,

(20·03) $$\int_{-A}^{A}\frac{|f(x)|^2}{1+x^2}\,dx$$

$$=\int_{-A}^{A}\frac{dx}{1+x^2}\frac{d}{dx}\int_{0}^{x}|f(\xi)|^2\,d\xi$$

$$=\frac{1}{1+A^2}\int_{0}^{A}|f(\xi)|^2\,d\xi-\frac{1}{1+A^2}\int_{0}^{-A}|f(\xi)|^2\,d\xi$$

$$-\int_{-A}^{A}\frac{d}{dx}\left(\frac{1}{1+x^2}\right)\left[\int_{0}^{x}|f(\xi)|^2\,d\xi\right]dx$$

$$=\frac{2A}{1+A^2}\frac{1}{2A}\int_{-A}^{A}|f(\xi)|^2\,d\xi+\int_{0}^{A}\frac{4x^2\,dx}{(1+x^2)^2}\frac{1}{2x}\int_{-x}^{x}|f(\xi)|^2\,d\xi$$

$$\leqslant\frac{2A}{1+A^2}\frac{1}{2A}\int_{-A}^{A}|f(\xi)|^2\,d\xi+4\int_{0}^{A}\frac{dx}{1+x^2}\frac{1}{2x}\int_{-x}^{x}|f(\xi)|^2\,d\xi$$

$$\leqslant\left\{\frac{2A}{1+A^2}+4\tan^{-1}A\right\}\limsup\frac{1}{2T}\int_{-T}^{T}|f(x)|^2\,dx$$

$$\leqslant(1+2\pi)\limsup\frac{1}{2T}\int_{-T}^{T}|f(x)|^2\,dx.$$

Thus if (20·01) is bounded, the function which is $f(x)/x$ for $|x|>1$ and 0 for $|x|\leqslant 1$ belongs to L_2, and, by the Plancherel theorem,

(20·04) $$s_1(u)=\underset{A\to\infty}{\text{l.i.m.}}\frac{1}{\sqrt{2\pi}}\left[\int_{1}^{A}+\int_{-A}^{-1}\right]\frac{f(x)\,e^{-iux}}{-ix}\,dx$$

exists. Furthermore, $f(x)$ is summable over $(-1,1)$, and

(20·05) $$s_2(u)=\frac{1}{\sqrt{2\pi}}\int_{-1}^{1}f(x)\frac{e^{-iux}-1}{-ix}\,dx$$

SPECIAL TAUBERIAN THEOREMS

exists for every u. Thus the function

(20·06) $$s(u) = s_1(u) + s_2(u)$$

exists for almost every u.

We have

(20·07) $s(u+\epsilon) - s(u-\epsilon)$
$$= \underset{A\to\infty}{\text{l.i.m.}} \frac{1}{\sqrt{2\pi}} \int_{-A}^{A} f(x) \frac{e^{ix\epsilon} - e^{-ix\epsilon}}{ix} e^{-iux} dx$$
$$= \underset{A\to\infty}{\text{l.i.m.}} \frac{1}{\sqrt{2\pi}} \int_{-A}^{A} f(x) \frac{2\sin\epsilon x}{x} e^{-iux} dx.$$

Thus $s(u+\epsilon) - s(u-\epsilon)$ is the Fourier transform of
$$f(x) \frac{2\sin\epsilon x}{x},$$
and, by the Plancherel theorem,

(20·08)
$$\frac{1}{4\pi\epsilon} \int_{-\infty}^{\infty} |s(u+\epsilon) - s(u-\epsilon)|^2 dx = \frac{1}{\pi\epsilon} \int_{-\infty}^{\infty} |f(x)|^2 \frac{\sin^2 \epsilon x}{x^2} dx.$$

The right-hand side of this expression may be regarded as a weighted average of $|f(x)|^2$, for
$$\frac{1}{\pi\epsilon} \int_{-\infty}^{\infty} \frac{\sin^2 \epsilon x}{x^2} dx = \frac{1}{\pi} \int_{-\infty}^{\infty} \frac{\sin^2 x}{x^2} dx = 1.$$

As $\epsilon \to 0$, the weighting becomes more nearly uniform over the interval $(-\infty, \infty)$. It becomes natural to inquire as to the relation between

(20·09) $$\lim_{T\to\infty} \frac{1}{2T} \int_{-T}^{T} |f(x)|^2 dx$$

and

(20·10) $$\lim_{\epsilon\to 0} \frac{1}{\pi\epsilon} \int_{-\infty}^{\infty} |f(x)|^2 \frac{\sin^2 \epsilon x}{x^2} dx.$$

We shall prove the Tauberian theorem:

THEOREM 21. *If $\phi(x) \geq 0$ for $0 \leq x < \infty$, and either of the limits*

(20·11) $$\lim_{T\to\infty} \frac{1}{T} \int_{0}^{T} \phi(x) dx$$

140 **SPECIAL TAUBERIAN THEOREMS**

or

(20·12) $$\lim_{\epsilon \to 0} \frac{2}{\pi\epsilon} \int_0^\infty \phi(x) \frac{\sin^2 \epsilon x}{x^2} dx$$

exists, then the other limit exists and assumes the same value.

Thus (20·09) and (20·10) are completely equivalent, and we have as a corollary (in view of 20·08):

THEOREM 22. *Let $f(x)$ be a function for which (20·01) is bounded in T. Let $s_1(u)$ be defined as in (20·04), $s_2(u)$ as in (20·05), and $s(u)$ as in (20·06). Then we shall have*

(20·13)
$$\lim_{\epsilon \to 0} \frac{1}{4\pi\epsilon} \int_{-\infty}^\infty |s(u+\epsilon) - s(u-\epsilon)|^2 du = \lim_{T \to \infty} \frac{1}{2T} \int_{-T}^T |f(x)|^2 dx,$$

in the sense that if either side of (20·13) exists, the other side exists and assumes the same value.

We now proceed to the proof of theorem 21. There will clearly be no restriction in supposing that $\phi(x) = 0$ for $x < 1$, inasmuch as over this region $\phi(x)/T$ and $2\phi(x)\sin^2\epsilon x/(\pi\epsilon x^2)$ tend monotonely to zero (the latter for sufficiently small values of ϵ), so that, by "monotone convergence,"

$$\lim_{T \to \infty} \frac{1}{T} \int_0^1 \phi(x) dx = \lim_{\epsilon \to 0} \frac{2}{\pi\epsilon} \int_0^1 \phi(x) \frac{\sin^2 \epsilon x}{x^2} dx = 0.$$

We shall accordingly suppose $\phi(x)$ to vanish over $(0, 1)$.

Let us now put

$$x = e^\xi; \quad T = e^\eta = \epsilon^{-1}; \quad \phi(x) = \psi(\xi).$$

The expression (20·11) becomes

(20·14) $$\lim_{\eta \to \infty} \int_{-\infty}^\eta e^{\xi - \eta} \psi(\xi) d\xi,$$

and (20·12) becomes

(20·15) $$\lim_{\eta \to \infty} \int_{-\infty}^\infty \frac{2}{\pi} \frac{\sin^2 e^{\xi - \eta}}{e^{\xi - \eta}} \psi(\xi) d\xi.$$

If either (20·14) or (20·15) is finite, we may conclude by a reasoning similar to the argument of (17·16) that

$$\varlimsup_{n \to \infty} \int_n^{n+1} \psi(\xi) d\xi < \infty.$$

SPECIAL TAUBERIAN THEOREMS

The function $\psi(\xi)$ is summable over any finite range, and vanishes for negative arguments. Thus, as in (17·17),
$$\int_n^{n+1} \psi(\xi)\,d\xi$$
is bounded. Hence if we put
$$g(\xi) = \int_0^\xi \psi(\eta)\,d\eta,$$
since ψ is non-negative, we have
$$\int_n^{n+1} |dg(\xi)| = g(n+1) - g(n) = \int_n^{n+1} \psi(\xi)\,d\xi < \text{const.}$$

We wish to show that the two propositions

(20·16) $\quad \lim\limits_{\xi \to \infty} \int_{-\infty}^\infty K_1(\xi - \eta)\,dg(\eta) = A \int_{-\infty}^\infty K_1(\xi)\,d\xi$

and

(20·17) $\quad \lim\limits_{\xi \to \infty} \int_{-\infty}^\infty K_2(\xi - \eta)\,dg(\eta) = A \int_{-\infty}^\infty K_2(\xi)\,d\xi$

are equivalent, where
$$K_1(\xi) = 0 \;\; (-\infty < \xi < 0); \quad K_1(\xi) = e^{-\xi} \;\; (0 < \xi < \infty)$$
and
$$K_2(\xi) = \frac{2}{\pi} e^\xi \sin^2 e^{-\xi}.$$

Let us notice that
$$\int_{-\infty}^\infty K_1(\xi)\,d\xi = \int_0^\infty e^{-\xi}\,d\xi = 1$$
and that
$$\int_{-\infty}^\infty K_2(\xi)\,d\xi = \frac{2}{\pi} \int_{-\infty}^\infty e^\xi \sin^2 e^{-\xi}\,d\xi$$
$$= \frac{2}{\pi} \int_0^\infty \frac{\sin^2 x}{x^2}\,dx = 1.$$

Thus, if (20·16) and (20·17) are equivalent, the two propositions

(20·18) $\quad \lim\limits_{\xi \to \infty} \int_{-\infty}^\infty K_1(\xi - \eta)\,dg(\eta) = A$

and

(20·19) $\quad \lim\limits_{\xi \to \infty} \int_{-\infty}^\infty K_2(\xi - \eta)\,dg(\eta) = A$

will be equivalent, and we shall have established theorem 21.

SPECIAL TAUBERIAN THEOREMS

To show the equivalence of (20·18) and (20·19), we invoke theorem 5. The kernel K_2 clearly belongs to M_1, and has a Fourier transform differing only by a constant factor from

$$\int_{-\infty}^{\infty} e^{\xi(1-iu)} \sin^2 e^{-\xi} d\xi$$

$$= \int_0^{\infty} x^{iu-2} \sin^2 x \, dx$$

$$= \int_0^{\infty} x^{iu-2} \left(\frac{1}{2} - \frac{e^{2ix}}{4} - \frac{e^{-2ix}}{4}\right) dx$$

$$= \lim_{\epsilon \to 0} \int_{\epsilon}^{\infty} x^{iu-2} \left(\frac{1}{2} - \frac{e^{2ix}}{4} - \frac{e^{-2ix}}{4}\right) dx$$

$$= \lim_{\epsilon \to 0} \left[\int_{\epsilon}^{\infty} x^{iu-2} \frac{1 - e^{2ix}}{4} dx + \int_{\epsilon}^{\infty} x^{iu-2} \frac{1 - e^{-2ix}}{4} dx \right].$$

In the first integral, we now deform the path of integration into the half-ordinate extending from ϵ to $\epsilon + i\infty$, and in the second, into the half-ordinate extending from ϵ to $\epsilon - i\infty$. The integrand will in each case be of the order of x^{-2} at infinity, so that the integral along the quadrant from a large positive value of x to a large imaginary value will tend to 0 as the radius becomes infinite, and the deformation of the path is legitimate. We have

$$\int_{-\infty}^{\infty} e^{\xi(1-iu)} \sin^2 e^{-\xi} d\xi$$

$$= \lim_{\epsilon \to 0} \left[i \int_0^{\infty} (\epsilon + iy)^{iu-2} \frac{1 - e^{2(i\epsilon-y)}}{4} dy \right.$$

$$\left. - i \int_0^{\infty} (\epsilon - iy)^{iu-2} \frac{1 - e^{2(-i\epsilon-y)}}{4} dy \right]$$

$$= \lim_{\epsilon \to 0} \int_0^{\infty} \left[i (\epsilon + iy)^{iu-2} \frac{1 - e^{2(i\epsilon-y)}}{4} \right.$$

$$\left. - i (\epsilon - iy)^{iu-2} \frac{1 - e^{2(-i\epsilon-y)}}{4} \right] dy.$$

Upon integration by parts, this becomes

(20·20)

$$\lim_{\epsilon \to 0} \left\{ \int_0^{\infty} \left[-\frac{(\epsilon + iy)^{iu-1}}{iu - 1} \frac{e^{2(i\epsilon-y)}}{2} - \frac{(\epsilon - iy)^{iu-1}}{iu - 1} \frac{e^{2(-i\epsilon-y)}}{2} \right] dy \right.$$

$$\left. - \frac{\epsilon^{iu-1}}{iu-1} \sin^2 \epsilon \right\},$$

SPECIAL TAUBERIAN THEOREMS

which, similarly,

$$= \lim_{\epsilon \to 0} \left\{ \int_0^\infty \left[\frac{(\epsilon+iy)^{iu}}{u(iu-1)} e^{2(i\epsilon-y)} - \frac{(\epsilon-iy)^{iu}}{u(iu-1)} e^{2(-i\epsilon-y)} \right] dy \right.$$
$$\left. + \frac{\epsilon^{iu}}{iu(iu-1)} \sin 2\epsilon - \frac{\epsilon^{iu-1}}{iu-1} \sin^2 \epsilon \right\}.$$

We may now proceed to the limit by the principle of dominated convergence, and (20·20) becomes

$$\int_0^\infty \frac{y^{iu}}{u(iu-1)} (i^{iu} - (-i)^{iu}) e^{-2y} dy$$
$$= \frac{\Gamma(iu+1)}{u(iu-1)} (e^{-\frac{\pi u}{2}} - e^{\frac{\pi u}{2}}) 2^{-iu-1},$$

which does not vanish for any real u.

The function K_1 is discontinuous, and hence does not belong to M_1. The functions

$$(20\cdot 21) \quad K_1(\xi, \epsilon) = \frac{1}{\epsilon} \int_\xi^{\xi+\epsilon} K_1(\eta) \, d\eta$$

$$\begin{cases} = 0; & (\xi < -\epsilon) \\ = \dfrac{1 - e^{-\xi-\epsilon}}{\epsilon}; & (-\epsilon \leqslant \xi < 0) \\ = e^{-\xi} \dfrac{1 - e^{-\epsilon}}{\epsilon}; & (0 \leqslant \xi) \end{cases}$$

do however belong to M_1. Their Fourier transforms are

$$\frac{1}{\sqrt{2\pi}} \int_{-\infty}^\infty e^{-iu\xi} \frac{d\xi}{\epsilon} \int_\xi^{\xi+\epsilon} K_1(\eta) \, d\eta$$
$$= \frac{1}{\sqrt{2\pi}} \int_{-\infty}^\infty e^{-iu\xi} \frac{d\xi}{\epsilon} \int_0^\epsilon K_1(\xi+\eta) \, d\eta$$
$$= \frac{1}{\epsilon \sqrt{2\pi}} \int_0^\epsilon d\eta \int_{-\eta}^\infty e^{-\xi(1+iu)-\eta} d\xi$$
$$= \frac{1}{\epsilon \sqrt{2\pi}} \int_0^\epsilon \frac{e^{iu\eta}}{1+iu} d\eta$$
$$= \frac{1}{\epsilon \sqrt{2\pi}} \frac{e^{iu\epsilon} - 1}{iu(1+iu)}.$$

144 SPECIAL TAUBERIAN THEOREMS

This vanishes only for $u = 2n\pi/\epsilon$ and there is hence no one u for which this vanishes for every ϵ. Thus, by theorem 7, (20·19) is equivalent to the statement that for all ϵ

(20·22)
$$\lim_{\xi \to \infty} \int_{-\infty}^{\infty} K_1(\xi - \eta, \epsilon)\, dg(\eta) = A \int_{-\infty}^{\infty} K_1(\xi, \epsilon)\, d\xi$$
$$= \frac{A}{\epsilon} \int_{-\infty}^{\infty} d\xi \int_{\xi}^{\xi+\epsilon} K_1(\eta)\, d\eta$$
$$= \frac{A}{\epsilon} \int_{-\infty}^{\infty} d\xi \int_{0}^{\epsilon} K_1(\xi + \eta)\, d\eta$$
$$= \frac{A}{\epsilon} \int_{0}^{\epsilon} d\eta \int_{-\infty}^{\infty} K_1(\xi)\, d\xi.$$

In lemma 7_5 no essential use was made of the continuity of $K(x)$. Thus we may establish in the same way that if (20·18) holds, (20·22) holds for every ϵ. Thus (20·18) implies (20·19). If we can now show that the validity of (20·22) for every ϵ implies that of (20·18), we shall have completed the proof of theorem 21 and hence of theorem 22.

To this end, let us notice that, by (20·21),

$$\frac{1 - e^{-\epsilon}}{\epsilon} K_1(\xi) \leqslant K_1(\xi, \epsilon) \leqslant \frac{e^{\epsilon} - 1}{\epsilon} K_1(\xi + \epsilon).$$

Thus, as $g(\eta)$ is monotone increasing,

$$\frac{1 - e^{-\epsilon}}{\epsilon} \overline{\lim_{\xi \to \infty}} \int_{-\infty}^{\infty} K_1(\xi - \eta)\, dg(\eta)$$
$$\leqslant \overline{\lim_{\xi \to \infty}} \int_{-\infty}^{\infty} K_1(\xi - \eta, \epsilon)\, dg(\eta) = A$$
$$\leqslant \frac{e^{\epsilon} - 1}{\epsilon} \overline{\lim_{\xi \to \infty}} \int_{-\infty}^{\infty} K_1(\xi - \eta + \epsilon)\, dg(\eta)$$
$$= \frac{e^{\epsilon} - 1}{\epsilon} \overline{\lim_{\xi \to \infty}} \int_{-\infty}^{\infty} K_1(\xi - \eta)\, dg(\eta).$$

As ϵ is arbitrarily small, we may make $(1 - e^{-\epsilon})/\epsilon$ and $(e^{\epsilon} - 1)/\epsilon$ as near as we wish to 1, and we have

$$\underline{\lim_{\xi \to \infty}} \int_{-\infty}^{\infty} K_1(\xi - \eta)\, dg(\eta) \leqslant A \leqslant \overline{\lim_{\xi \to \infty}} \int_{-\infty}^{\infty} K_1(\xi - \eta)\, dg(\eta),$$

SPECIAL TAUBERIAN THEOREMS 145

which is equivalent to (20·18). This completes the proof of theorem 21.

A closely related theorem is:

THEOREM 23. *Let $\phi(x)$ be a bounded function, and let either of the limits*

(20·23) $$\lim_{\epsilon \to 0} \frac{1}{\epsilon} \int_0^\epsilon \phi(x)\, dx$$

or

(20·24) $$\lim_{T \to \infty} \frac{2}{\pi T} \int_0^\infty \phi(x) \frac{\sin^2 Tx}{x^2}\, dx$$

exist. Then the other limit exists and assumes the same value.

In this theorem the rôles of 0 and ∞ are reversed in comparison with theorem 21. We put

$$x = e^{-\xi}; \quad T = e^\eta = \epsilon^{-1}; \quad \phi(x) = \psi(\xi).$$

The expression (20·23) becomes

(20·25) $$\lim_{\eta \to \infty} \int_\eta^\infty e^{\eta-\xi} \psi(\xi)\, d\xi,$$

and (20·24) becomes

(20·26) $$\lim_{\eta \to \infty} \int_{-\infty}^\infty \frac{2}{\pi} \frac{\sin^2 e^{\eta-\xi}}{e^{\eta-\xi}} \psi(\xi)\, d\xi.$$

As we have assumed ψ to be bounded, we may apply theorem 4 to show the complete equivalence of (20·25) and (20·26). We have only to show that neither

$$\int_{-\infty}^0 e^{\xi(1-iu)}\, d\xi$$

nor

$$\int_{-\infty}^\infty e^{-\xi(1+iu)} \sin^2 e^\xi\, d\xi$$

vanishes for any real u. The first is $1/(1-iu)$ and does not vanish, while the second is equal to

$$\int_{-\infty}^\infty e^{\xi(1+iu)} \sin^2 e^{-\xi}\, d\xi,$$

and this, by (20·19, 20), is equal to

$$\frac{\Gamma(1-iu)(e^{\pi u/2} - e^{-\pi u/2}) 2^{iu-1}}{u(iu+1)},$$

which vanishes for no real u. Thus theorem 23 is established.

A corollary of theorem 22 is:

THEOREM 24. *Let $f(x)$ be a function for which (20·01) is bounded in T. Let $s_1(u)$ be defined as in (20·04), $s_2(u)$ as in (20·05), and $s(u)$ as in (20·06). Let $s(u)$ be of limited total variation over $(-\infty, \infty)$. Then*

(20·261) $$\lim_{T\to\infty} \frac{1}{2T} \int_{-T}^{T} |f(x)|^2 \, dx$$

exists and is equal to $1/(2\pi)$ times the sum of the squares of the jumps of $s(u)$ at all its points of discontinuity. In particular, if $s(u)$ is continuous, the expression (20·261) is equal to 0.

We may write
$$s(u) = s_\eta(u) + t_\eta(u),$$
where $t_\eta(u)$ is a step-function constant except at the jumps of $s(u)$ exceeding η in magnitude, and there having the same jump as $s(u)$. The function $s_\eta(u)$ will have no jump exceeding η in magnitude, and both $s_\eta(u)$ and $t_\eta(u)$ will have total variations not exceeding V, the total variation of $s(u)$.

By a form of the Minkowski inequality,

$$\left| \left\{ \frac{1}{2\epsilon} \int_{-\infty}^{\infty} |s(u+\epsilon) - s(u-\epsilon)|^2 \, du \right\}^{\frac{1}{2}} \right.$$
$$- \left\{ \frac{1}{2\epsilon} \int_{-\infty}^{\infty} |t_\eta(u+\epsilon) - t_\eta(u-\epsilon)|^2 \, du \right\}^{\frac{1}{2}} \right|$$
$$\leq \left\{ \frac{1}{2\epsilon} \int_{-\infty}^{\infty} |s_\eta(u+\epsilon) - s_\eta(u-\epsilon)|^2 \, du \right\}^{\frac{1}{2}}.$$

Now, if ϵ is small enough,

$$\frac{1}{2\epsilon} \int_{-\infty}^{\infty} |t_\eta(u+\epsilon) - t_\eta(u-\epsilon)|^2 \, du$$

will be the sum of the squares of all the jumps of $s(u)$ exceeding η in magnitude. To see this, let us notice that

$$|t_\eta(u+\epsilon) - t_\eta(u-\epsilon)|$$

will vanish except in regions of length 2ϵ extending a distance ϵ on each side of each jump, in which it will be equal to the absolute magnitude of the jump.

SPECIAL TAUBERIAN THEOREMS

The expression $|s_\eta(u+\epsilon) - s_\eta(u-\epsilon)|$ can be made less than 2η by choosing ϵ small enough. Over any finite range of u, this choice of ϵ may be made independently of u. To show this, we appeal to the Heine-Borel theorem, and to the fact that about each point u we can place an interval in which the fluctuation of $s_\eta(u)$ is less than 2η. Hence we can cover any finite range of u by a finite set of overlapping intervals in which the fluctuation of $s_\eta(u)$ is less than 2η, and if we choose ϵ less than half the overlap of any two of these intervals,

$$|s_\eta(u+\epsilon) - s_\eta(u-\epsilon)|$$

will be less than 2η for any u in the finite range.

It follows that by taking ϵ small enough, we shall have

$$(20\cdot 265) \qquad \frac{1}{2\epsilon} \int_{-\infty}^{\infty} |s_\eta(u+\epsilon) - s_\eta(u-\epsilon)|^2 \, du < 3\eta V.$$

For let us put

$$\frac{1}{2\epsilon} \int_{-\infty}^{\infty} |s_\eta(u+\epsilon) - s_\eta(u-\epsilon)|^2 \, du$$

$$= \frac{1}{2\epsilon} \int_{-\infty}^{-A} |s_\eta(u+\epsilon) - s_\eta(u-\epsilon)|^2 \, du$$

$$+ \frac{1}{2\epsilon} \int_{-A}^{A} |s_\eta(u+\epsilon) - s_\eta(u-\epsilon)|^2 \, du$$

$$+ \frac{1}{2\epsilon} \int_{A}^{\infty} |s_\eta(u+\epsilon) - s_\eta(u-\epsilon)|^2 \, du.$$

The function $|s_\eta|$ will have some finite upper bound, say S. Then

$$\frac{1}{2\epsilon} \int_{A}^{\infty} |s_\eta(u+\epsilon) - s_\eta(u-\epsilon)|^2 \, du$$

$$\leqslant 2S \frac{1}{2\epsilon} \int_{A}^{\infty} |s_\eta(u+\epsilon) - s_\eta(u-\epsilon)| \, du$$

$$= 2S \frac{1}{2\epsilon} \left[\int_{A}^{A+2\epsilon} + \int_{A+2\epsilon}^{A+4\epsilon} + \dots \right] |s_\eta(u+\epsilon) - s_\eta(u-\epsilon)| \, du$$

$$= 2S \frac{1}{2\epsilon} \int_{0}^{2\epsilon} \{ |s_\eta(u+A+\epsilon) - s_\eta(u+A-\epsilon)|$$
$$+ |s_\eta(u+A+3\epsilon) - s_\eta(u+A+\epsilon)|$$
$$+ |s_\eta(u+A+5\epsilon) - s_\eta(u+A+3\epsilon)|$$
$$+ \dots \}.$$

148   SPECIAL TAUBERIAN THEOREMS

This cannot exceed $2S$ multiplied by the total variation of s from $A - \epsilon$ to infinity, and can be made uniformly small in both ϵ and η by taking A large enough. The same applies to the integral from $-\infty$ to $-A$. Let us make these two integrals together less than ηV. There only remains the integral from $-A$ to A. This cannot exceed

$$\max_{-A<u<A} |s_\eta(u+\epsilon) - s_\eta(u-\epsilon)| \frac{1}{2\epsilon} \int_{-A}^{A} |s_\eta(u+\epsilon) - s_\eta(u-\epsilon)|\, du,$$

and we may take ϵ so small that this is less than

$$2\eta \frac{1}{2\epsilon} \int_{-\infty}^{\infty} |s_\eta(u+\epsilon) - s_\eta(u-\epsilon)|\, du$$

$$= 2\eta \frac{1}{2\epsilon} \int_{-\epsilon}^{\epsilon} \sum_{n=-\infty}^{\infty} |s_\eta(u+(2n+1)\epsilon) - s_\eta(u+(2n-1)\epsilon)|\, du$$

$$\leqslant 2\eta \frac{1}{2\epsilon} \int_{-\epsilon}^{\epsilon} V\, du \leqslant 2\eta V.$$

This establishes (20·265).

It follows from (20·265) that if we choose ϵ small enough,

$$\left\{ \frac{1}{2\epsilon} \int_{-\infty}^{\infty} |s(u+\epsilon) - s(u-\epsilon)|^2\, du \right\}^{\frac{1}{2}}$$

differs from the square root of the sum of the squares of those jumps of $s(u)$ exceeding η in magnitude by less than $(3\eta V)^{\frac{1}{2}}$. If we call this sum Σ_η, we have

$$\overline{\lim_{\epsilon \to 0}} \left\{ \frac{1}{2\epsilon} \int_{-\infty}^{\infty} |s(u+\epsilon) - s(u-\epsilon)|^2\, du \right\}^{\frac{1}{2}} < \{\Sigma_\eta\}^{\frac{1}{2}} + (3\eta V)^{\frac{1}{2}},$$

$$\underline{\lim_{\epsilon \to 0}} \left\{ \frac{1}{2\epsilon} \int_{-\infty}^{\infty} |s(u+\epsilon) - s(u-\epsilon)|^2\, du \right\}^{\frac{1}{2}} > \{\Sigma_\eta\}^{\frac{1}{2}} - (3\eta V)^{\frac{1}{2}}.$$

It follows that

$$\overline{\lim_{\epsilon \to 0}} \left\{ \frac{1}{2\epsilon} \int_{-\infty}^{\infty} |s(u+\epsilon) - s(u-\epsilon)|^2\, du \right\}^{\frac{1}{2}}$$

$$- \underline{\lim_{\epsilon \to 0}} \left\{ \frac{1}{2\epsilon} \int_{-\infty}^{\infty} |s(u+\epsilon) - s(u-\epsilon)|^2\, du \right\}^{\frac{1}{2}} < 2(3\eta V)^{\frac{1}{2}},$$

and since η is arbitrarily small,

$$\lim_{\epsilon \to 0} \left\{ \frac{1}{2\epsilon} \int_{-\infty}^{\infty} |s(u+\epsilon) - s(u-\epsilon)|^2\, du \right\}^{\frac{1}{2}}$$

SPECIAL TAUBERIAN THEOREMS

exists, and differs from $\{\Sigma_\eta\}^{\frac{1}{2}}$ by less than $(3\eta V)^{\frac{1}{2}}$. Hence

$$\lim_{\eta \to 0} \{\Sigma_\eta\}^{\frac{1}{2}} = \lim_{\epsilon \to 0} \left\{ \frac{1}{2\epsilon} \int_{-\infty}^{\infty} |s(u+\epsilon) - s(u-\epsilon)|^2 du \right\}^{\frac{1}{2}},$$

and

(20·27) $\qquad \lim_{\eta \to 0} \Sigma_\eta = \lim_{\epsilon \to 0} \frac{1}{2\epsilon} \int_{-\infty}^{\infty} |s(u+\epsilon) - s(u-\epsilon)|^2 du.$

The left-hand side of (20·27) is however the sum of the squares of *all* the jumps of $s(u)$. Theorem 22 completes the proof of theorem 24. Of course, if $s(u)$ is of limited total variation and has no jumps,

$$\lim_{\epsilon \to 0} \frac{1}{2\epsilon} \int_{-\infty}^{\infty} |s(u+\epsilon) - s(u-\epsilon)|^2 du = 0,$$

and hence $\qquad \lim_{T \to \infty} \frac{1}{2T} \int_{-T}^{T} |f(x)|^2 dx = 0.$

CHAPTER IV

GENERALIZED HARMONIC ANALYSIS

§ 21. The Spectrum of a Function.

Let $f(x) = \sum_{1}^{n} A_j e^{i\Lambda_j x}$ be a given trigonometrical polynomial. Then

(21·01) $\quad f(x+\xi)\bar{f}(\xi) = \sum_{j=1}^{n}\sum_{k=1}^{n} A_j \bar{A}_k e^{i\Lambda_j x} e^{i(\Lambda_j - \Lambda_k)\xi}.$

Now, if $\mu \neq 0$,

(21·02) $\quad \lim_{T\to\infty} \frac{1}{2T}\int_{-T}^{T} e^{i\mu\xi} d\xi = \lim_{T\to\infty} \frac{\sin \mu T}{\mu T} = 0.$

On the other hand,

(21·03) $\quad \lim_{T\to\infty} \frac{1}{2T}\int_{-T}^{T} e^{i0\xi} d\xi = 1.$

Thus

$$\lim_{T\to\infty} \frac{1}{2T}\int_{-T}^{T} f(x+\xi)\bar{f}(\xi) d\xi = \sum_{1}^{n} A_j \bar{A}_j e^{i\Lambda_j x} = \sum_{1}^{n} |A_j|^2 e^{i\Lambda_j x}.$$

In other words, if we put

(21·04) $\quad \phi(x) = \lim_{T\to\infty} \frac{1}{2T}\int_{-T}^{T} f(x+\xi)\bar{f}(\xi) d\xi,$

then $\phi(x)$ exists for every x, is continuous, and consists of terms with the same frequency as those constituting $f(x)$, with amplitudes the square of the amplitudes of the corresponding terms of $f(x)$, and all with such a phase that they are real for $x = 0$.

Let us now consider more general classes of functions $f(x)$ than the trigonometric polynomials considered above. In case f is measurable over $(-\infty, \infty)$ and $\phi(x)$ as defined in (21·04) exists for every x, we shall say that $f(x)$ belongs to class S. In case $f(x)$ belongs to S and $\phi(x)$ is continuous, we shall say that $f(x)$ belongs to S'. Thus every trigonometrical polynomial belongs to S'. We shall give several examples of functions $f(x)$ belonging to S' or S.

(i) $f(x)$ belongs to L_2. Then $\phi(x) \equiv 0$, and $f(x)$ belongs to S'.

(ii) $f(x) = e^{i|x|^{\frac{1}{2}}}$. Then

$$\phi(x) = \lim_{T \to \infty} \frac{1}{2T} \int_{-T}^{T} \exp i(|x+\xi|^{\frac{1}{2}} - |\xi|^{\frac{1}{2}}) d\xi$$

$$= \lim_{T \to \infty} \frac{1}{2T} \int_{-T}^{T} \exp i \frac{|x+\xi| - |\xi|}{|x+\xi|^{\frac{1}{2}} + |\xi|^{\frac{1}{2}}} d\xi$$

$$= \lim_{T \to \infty} \frac{1}{2T} \int_{-T}^{T} (1 + O(|\xi|^{-\frac{1}{2}})) d\xi$$

$$= 1.$$

Thus $f(x)$ belongs to S'.

(iii) $f(x) = e^{ix^2}$. Then

$$\phi(x) = \lim_{T \to \infty} \frac{1}{2T} \int_{-T}^{T} \exp i((x+\xi)^2 - \xi^2) d\xi$$

$$= \lim_{T \to \infty} \frac{e^{ix^2}}{2T} \int_{-T}^{T} e^{2ix\xi} d\xi$$

$$= \begin{cases} 1 & \text{if } x = 0; \\ 0 & \text{if } x \neq 0; \end{cases}$$

in accordance with (21·03) and (21·02). Thus $f(x)$ belongs to S but not to S'.

(iv) λ is a number in $(0, 1)$, with the binary expansion

$$0 \cdot a_1 a_2 a_3 a_4 \ldots$$

The function $f(x)$ is defined as $2a_{2n+1} - 1$ for $n < x \leqslant n+1$, if n is a positive integer or 0, and as $2a_{2n} - 1$ if $-n < x \leqslant 1-n$. We wish to show that for *almost all* values of λ,

(21·05) $\quad \phi(x) = 1 - |x| \quad (|x| \leqslant 1); \quad \phi(x) = 0 \quad (x > 1).$

It will follow that for these values of λ, $f(x)$ will belong to S'.

What we purpose to show may also be stated as follows: if $f(x)$ has over each interval $(n, n+1)$ $(-\infty < n < \infty)$ either the value $+1$ or the value -1, and if each choice of these values is independent of all the others, then the probability that $\phi(x)$ will not have the value given in (21·05) will be 0. The necessary reduction of probabilities to Lebesgue measure has been made by Borel* and Steinhaus†.

To begin with, let us note that if $0 \leqslant n \leqslant |x| \leqslant n+1$, we shall have

(21·06)
$$\phi(x) = \lim_{N\to\infty} \frac{1}{2N} \int_{-N}^{N} f(x+\xi)\bar{f}(\xi)\,d\xi$$
$$= \lim_{N\to\infty} \frac{1}{2N} \sum_{m=-N}^{N-1} \int_{m}^{m+1} f(x+\xi)f(m+1)\,d\xi$$
$$= \lim_{N\to\infty} \frac{1}{2N} \sum_{m=-N}^{N-1} \left\{ \int_{m}^{m+n+1-x} f(m+n+1)f(m+1)\,d\xi \right.$$
$$\left. + \int_{m+n+1-x}^{m+1} f(m+n+2)f(m+1)\,d\xi \right\}$$
$$= \lim_{N\to\infty} \frac{1}{2N} \sum_{m=-N}^{N} \{(n+1-x)f(m+n+1)f(m+1)$$
$$+ (x-n)f(m+n+2)f(m+1)\}$$
$$= (n+1-x)\phi(n) + (x-n)\phi(n+1).$$

Formula (21·06) is to be taken in the sense that if the last term exists, $\phi(x)$ will exist and assume the same value. The replacement of the general variable T in (21·04) by the integral variable N in the first line of (21·06) is readily seen to be no real restriction.

In view of (21·06), all that we need prove to establish (21·05) is that except for a set of values of λ of measure zero, $\phi(0) = 1$ and $\phi(n) = 0$ when n is an integer other than 0. That $\phi(0) = 1$ is an immediate consequence of the fact that $|f(x)| = 1$. If now we can show for each particular non-zero integer n that $\phi(n) = 0$ except for a null set of values of λ, we may appeal to the theorem that the logical sum of a denumerable set of null sets is null to complete the proof of (21·05).

* E. Borel, "Les probabilités dénombrables et leurs applications arithmétiques," *Rend. di Palermo*, 27 (1909), 247–271.

† H. Steinhaus, "Les probabilités dénombrables et leur rapport à la théorie de la mesure," *Fund. Math.* 4 (1923), 286–318.

GENERALIZED HARMONIC ANALYSIS

Let us now consider $f(m+n)f(m)$ for fixed n and variable m. For any m it assumes either the value $+1$ or the value -1, and if we take any finite consecutive set of numbers m, any sequence of signs is as probable as any other—that is, any sequence of signs corresponds to a region of λ of the same Lebesgue measure as that corresponding to any other. If we take $2N$ consecutive values of $f(m+n)f(m)$, the measure of all those regions of λ for which the sum of these values exceeds in modulus $N\epsilon$ will not exceed

$$2^{2N+1} \sum_{k=\left[\frac{N\epsilon}{2}\right]}^{N} \frac{(2N)!}{(N-k)!(N+k)!}$$

by the binomial theorem and ordinary considerations from the theory of probabilities. By Stirling's theorem this is asymptotically

$$2 \sum_{k=\left[\frac{N\epsilon}{2}\right]}^{N} \frac{1}{\left(1-\frac{k}{N}\right)^{N-k}\left(1+\frac{k}{N}\right)^{N+k}} \left(\frac{N}{\pi(N^2-k^2)}\right)^{\frac{1}{2}}$$

$$\sim \sum_{k=\left[\frac{N\epsilon}{2}\right]}^{N} 2 e^{\frac{k}{N}(N-k-N-k)} \left(\frac{N}{\pi(N^2-k^2)}\right)^{\frac{1}{2}}$$

$$= O(N^{\frac{1}{2}} e^{-\frac{N\epsilon^2}{2}}).$$

Inasmuch as $\sum_{N=1}^{\infty} N^{\frac{1}{2}} e^{-\frac{N\epsilon^2}{2}}$

converges, the chance is zero that there should fail to be an integral value of N such that, from that value on,

$$\sum_{m=-N}^{N} f(m+n)f(m) \leqslant N\epsilon.$$

Thus, except for a set of values of λ of zero measure,

$$\overline{\lim_{N\to\infty}} \left|\frac{1}{2N}\int_{-N}^{N} f(n+\xi)\bar{f}(\xi)\,d\xi\right| \leqslant \epsilon/2,$$

and since ϵ is arbitrary, and the sum of a denumerable set of null sets is null, we shall have

$$\lim_{N\to\infty} \frac{1}{2N}\int_{-N}^{N} f(n+\xi)\bar{f}(\xi)\,d\xi = 0$$

154   GENERALIZED HARMONIC ANALYSIS

except for a null set of values of λ. The transition from integral to general values of N offers no difficulty, and we have
$$\phi(n) = 0,$$
thus completing the proof of (21·05).

By the Schwarz inequality, we have:

THEOREM 25. *If $f(x)$ belongs to S, and y is any real number,*
(21·07) $$|\phi(y)| \leqslant \phi(0).$$
For
$$|\phi(y)| = \left| \lim_{T \to \infty} \frac{1}{2T} \int_{-T}^{T} f(y+\xi) \bar{f}(\xi) \, d\xi \right|$$
$$= \lim_{T \to \infty} \left| \frac{1}{2T} \int_{-T}^{T} f(y+\xi) \bar{f}(\xi) \, d\xi \right|$$
$$\leqslant \overline{\lim_{T \to \infty}} \frac{1}{2T} \left\{ \int_{-T}^{T} |f(y+\xi)|^2 \, d\xi \int_{-T}^{T} |f(\xi)|^2 \, d\xi \right\}^{\frac{1}{2}}$$
$$\leqslant \left\{ \overline{\lim_{T \to \infty}} \frac{1}{2T} \int_{-T+y}^{T+y} |f(\xi)|^2 \, d\xi \; \overline{\lim_{T \to \infty}} \frac{1}{2T} \int_{-T}^{T} |f(\xi)|^2 \, d\xi \right\}^{\frac{1}{2}}$$
$$\leqslant \left\{ \overline{\lim_{T \to \infty}} \frac{1}{2T} \int_{-T-|y|}^{T+|y|} |f(\xi)|^2 \, d\xi \; \overline{\lim_{T \to \infty}} \frac{1}{2T} \int_{-T}^{T} |f(\xi)|^2 \, d\xi \right\}^{\frac{1}{2}}$$
$$\leqslant \left\{ \overline{\lim_{T \to \infty}} \frac{T+|y|}{T} \; \overline{\lim_{T \to \infty}} \frac{1}{2(T+|y|)} \right.$$
$$\left. \times \int_{-T-|y|}^{T+|y|} |f(\xi)|^2 \, d\xi \; \overline{\lim_{T \to \infty}} \frac{1}{2T} \int_{-T}^{T} |f(\xi)|^2 \, d\xi \right\}^{\frac{1}{2}}$$
$$= \{1 \cdot \phi(0) \cdot \phi(0)\}^{\frac{1}{2}} = \phi(0).$$

Thus every $\phi(x)$ pertaining to a function $f(x)$ belonging to S is bounded.

By another application of the Schwarz inequality, we have:

THEOREM 26. *If $f(x)$ belongs to S and $\phi(x)$ is continuous at 0, then it is continuous for all real arguments and $f(x)$ belongs to S'.*

For
(21·08) $$|\phi(x+\epsilon) - \phi(x)|$$
$$= \lim_{T \to \infty} \left| \frac{1}{2T} \int_{-T}^{T} \{f(x+\xi+\epsilon) - f(x+\xi)\} \bar{f}(\xi) \, d\xi \right|$$
$$\leqslant \overline{\lim_{T \to \infty}} \frac{1}{2T} \left\{ \int_{-T}^{T} |f(x+\xi+\epsilon) - f(x+\xi)|^2 \, d\xi \int_{-T}^{T} |f(\xi)|^2 \, d\xi \right\}^{\frac{1}{2}}.$$

GENERALIZED HARMONIC ANALYSIS

We now appeal to the following lemma:

Lemma 26$_1$. *If $\psi(x)$ is a positive function, a is any real constant, and if*

$$\lim_{T \to \infty} \frac{1}{2T} \int_{-T}^{T} \psi(x)\,dx = A,$$

then
$$\lim_{T \to \infty} \frac{1}{2T} \int_{-T+a}^{T+a} \psi(x)\,dx = A.$$

To see this, let us notice that if $T > |a|$,

$$\left(1 - \frac{|a|}{T}\right) \frac{1}{2(T-|a|)} \int_{-T+|a|}^{T-|a|} \psi(x)\,dx \leq \frac{1}{2T} \int_{-T+a}^{T+a} \psi(x)\,dx$$
$$\leq \left(1 + \frac{|a|}{T}\right) \frac{1}{2(T+|a|)} \int_{-T+|a|}^{T+|a|} \psi(x)\,dx,$$

and that both the first and the last of these three expressions tend to A with increasing T. In particular, if $\phi(0)$ exists,

$$\lim_{T \to \infty} \frac{1}{2T} \int_{-T}^{T} |f(\xi+a)|^2\,d\xi = \lim_{T \to \infty} \frac{1}{2T} \int_{-T+a}^{T+a} |f(\xi)|^2\,d\xi$$
$$= \lim_{T \to \infty} \frac{1}{2T} \int_{-T}^{T} |f(\xi)|^2\,d\xi = \phi(0).$$

Again, if $f(x)$ belongs to S,

(21·09)
$$\phi(y) = \lim_{T \to \infty} \frac{1}{2T} \int_{-T}^{T} f(\xi+y) \bar{f}(\xi)\,d\xi$$
$$= \frac{1}{4} \lim_{T \to \infty} \frac{1}{2T} \int_{-T}^{T} \{|f(\xi+y)+f(\xi)|^2 - |f(\xi+y)-f(\xi)|^2$$
$$+ i|f(\xi+y)+if(\xi)|^2 - i|f(\xi+y)-if(\xi)|^2\}\,d\xi.$$

This latter limit may be resolved into a linear combination of four means of positive quantities, each of which exists independently. Thus, by lemma 26$_1$,

$$\phi(y) = \frac{1}{4} \lim_{T \to \infty} \frac{1}{2T} \int_{-T+a}^{T+a} \{|f(\xi+y)+f(\xi)|^2 - |f(\xi+y)-f(\xi)|^2$$
$$+ i|f(\xi+y)+if(\xi)|^2 - i|f(\xi+y)-if(\xi)|^2\}\,d\xi$$
$$= \lim_{T \to \infty} \frac{1}{2T} \int_{-T+a}^{T+a} f(\xi+y)\bar{f}(\xi)\,d\xi.$$

We thus have

$$\lim_{T\to\infty} \frac{1}{2T}\int_{-T}^{T} |f(x+\xi+\epsilon) - f(x+\xi)|^2 \, d\xi$$
$$= \lim_{T\to\infty} \frac{1}{2T}\int_{-T}^{T} |f(x+\xi+\epsilon)|^2 \, d\xi + \lim_{T\to\infty} \frac{1}{2T}\int_{-T}^{T} |f(x+\xi)|^2 \, d\xi$$
$$- \lim_{T\to\infty} \frac{1}{2T}\int_{-T}^{T} f(x+\xi+\epsilon)\bar{f}(x+\xi) \, d\xi$$
$$- \lim_{T\to\infty} \frac{1}{2T}\int_{-T}^{T} f(x+\xi)\bar{f}(x+\xi+\epsilon) \, d\xi$$
$$= 2\phi(0) - \phi(\epsilon) - \phi(-\epsilon).$$

Thus (21·08) yields us

$$|\phi(x+\epsilon) - \phi(x)| \leq \{\phi(0)[2\phi(0) - \phi(\epsilon) - \phi(-\epsilon)]\}^{\frac{1}{2}},$$

from which theorem 26 follows immediately.

Let us return to (21·09). If we define $s(u)$ as in (20·04–20·06), the function bearing to $f(\xi+y)$ the same relation that $s(u)$ does to $f(\xi)$ will be

$$s_y(u) = \frac{1}{\sqrt{2\pi}} \left\{ \underset{A\to\infty}{\text{l.i.m.}} \left[\int_1^A + \int_{-A}^{-1}\right] \frac{f(\xi+y)e^{-iu\xi}}{-i\xi} \, d\xi \right.$$
$$\left. + \int_{-1}^{1} f(\xi+y) \frac{e^{-iu\xi}-1}{-i\xi} \, d\xi \right\}.$$

We shall have

(21·10) $s_y(u+\epsilon) - s_y(u-\epsilon) - e^{iuy}(s(u+\epsilon) - s(u-\epsilon))$

$$= \frac{1}{\sqrt{2\pi}} \left\{ \underset{A\to\infty}{\text{l.i.m.}} \int_{-A}^{A} f(\xi+y) \frac{2\sin\xi\epsilon}{\xi} e^{-iu\xi} d\xi \right.$$
$$\left. - \underset{A\to\infty}{\text{l.i.m.}} \int_{-A}^{A} f(\xi) \frac{2\sin\xi\epsilon}{\xi} e^{-iu(\xi-y)} d\xi \right\}.$$

Let us notice that in the definition of the Fourier transform we might replace the symbol $\underset{A\to\infty}{\text{l.i.m.}} \int_{-A}^{A}$ by $\underset{A,B\to\infty}{\text{l.i.m.}} \int_{-A}^{B}$ without any restriction or generalization (cf. (9·02)), so that (21·10) becomes

$$\underset{A\to\infty}{\text{l.i.m.}} \frac{1}{\sqrt{2\pi}} \int_{-A}^{A} f(\xi) \left[\frac{2\sin(\xi-y)\epsilon}{\xi-y} - \frac{2\sin\xi\epsilon}{\xi}\right] e^{-iu(\xi-y)} d\xi.$$

GENERALIZED HARMONIC ANALYSIS

Thus, by the Plancherel theorem,

(21·11) $\int_{-\infty}^{\infty} |s_y(u+\epsilon) - s_y(u-\epsilon) - e^{iuy}(s(u+\epsilon) - s(u-\epsilon))|^2 du$

$$= \int_{-\infty}^{\infty} |f(\xi)|^2 \left[\frac{2\sin(\xi-y)\epsilon}{\xi-y} - \frac{2\sin\xi\epsilon}{\xi} \right]^2 d\xi.$$

Now,

(21·12) $\left| \frac{2\sin(\xi-y)\epsilon}{\xi-y} - \frac{2\sin\xi\epsilon}{\xi} \right| = \left| 2\int_{\xi-y}^{\xi} \frac{d}{dw}\left(\frac{\sin\epsilon w}{w}\right) dw \right|$

$$= 2\left| \int_{\xi-y}^{\xi} \frac{\epsilon w \cos\epsilon w - \sin\epsilon w}{w^2} dw \right|$$

$$\leqslant 2\epsilon^2 |y| \max_{w \text{ in } (\xi-y,\xi)} \left| \frac{\cos\epsilon w}{\epsilon w} - \frac{\sin\epsilon w}{\epsilon^2 w^2} \right|$$

$$\leqslant 4\epsilon^2 |y| \max \frac{1}{|\epsilon w|}.$$

Also,

(21·13) $\left| \frac{2\sin(\xi-y)\epsilon}{\xi-y} - \frac{2\sin\xi\epsilon}{\xi} \right| \leqslant 4\epsilon.$

Let us consider the various possible relations of ξ, y, and $\xi-y$, and let us put

(21·14) $\left| \frac{2\sin(\xi-y)\epsilon}{\xi-y} - \frac{2\sin\xi\epsilon}{\xi} \right| = T.$

In case $0 \leqslant \xi/2 \leqslant \xi-y \leqslant \xi$, we have, by (21·12),

$$T \leqslant \frac{8\epsilon|y|}{|\xi|} \leqslant \frac{16\epsilon|y|}{|y|+|\xi|}.$$

In case $0 \leqslant \xi-y \leqslant \xi/2 \leqslant \xi$, we have, by (21·13),

$$T \leqslant 4\epsilon \leqslant \frac{16\epsilon|y|}{|y|+|\xi|}.$$

In case $0 \leqslant \xi \leqslant \xi-y < \frac{3\xi}{2}$, we have, by (21·12),

$$T \leqslant \frac{4\epsilon|y|}{|\xi|} \leqslant \frac{16\epsilon|y|}{|y|+|\xi|}.$$

In case $0 \leqslant \xi$, $3\xi/2 \leqslant \xi-y$, we have, by (21·13),

$$T \leqslant 4\epsilon \leqslant \frac{16\epsilon|y|}{|\xi|+|y|}.$$

158 GENERALIZED HARMONIC ANALYSIS

Thus $T \leqslant 16\epsilon\,|y|/(|\xi|+|y|)$ whenever ξ and $\xi-y$ are both $\geqslant 0$. This result will also hold when both are $\leqslant 0$. In case they lie on opposite sides of 0, we have $|\xi| < |y|$, and

$$T \leqslant 4\epsilon \leqslant \frac{16\epsilon\,|y|}{|\xi|+|y|}.$$

Thus, in all instances,

(21·15) $$T \leqslant \frac{16\epsilon\,|y|}{|\xi|+|y|},$$

and, by theorem 20 and (21·11),

$$\int_{-\infty}^{\infty} |s_y(u+\epsilon) - s_y(u-\epsilon) - e^{iuy}(s(u+\epsilon)-s(u-\epsilon))|^2\,du = O(\epsilon^2).$$

Thus, by theorem 22 and the Minkowski inequality, if $|w|=1$,

$$\left\{\lim_{B\to\infty}\frac{1}{2B}\int_{-B}^{B}|f(\xi+y)+wf(\xi)|^2\,d\xi\right\}^{\frac{1}{2}}$$

$$= \lim_{\epsilon\to 0}\left\{\frac{1}{4\pi\epsilon}\int_{-\infty}^{\infty}|s_y(u+\epsilon)-s_y(u-\epsilon)\right.$$

$$\left.+w(s(u+\epsilon)-s(u-\epsilon))|^2\,du\right\}^{\frac{1}{2}}$$

$$= \lim_{\epsilon\to 0}\left\{O(\epsilon^{\frac{1}{2}}) + \left\{\frac{1}{4\pi\epsilon}\int_{-\infty}^{\infty}|e^{iuy}+w|^2\right.\right.$$

$$\left.\left.\times |s(u+\epsilon)-s(u-\epsilon)|^2\,du\right\}^{\frac{1}{2}}\right\}$$

$$= \lim_{\epsilon\to 0}\left\{\frac{1}{4\pi\epsilon}\int_{-\infty}^{\infty}(2+we^{-iuy}+\overline{w}e^{iuy})\right.$$

$$\left.\times |s(u+\epsilon)-s(u-\epsilon)|^2\,du\right\}^{\frac{1}{2}}.$$

Hence

(21·16) $$\lim_{B\to\infty}\frac{1}{2B}\int_{-B}^{B}|f(\xi+y)+wf(\xi)|^2\,d\xi$$

$$= \lim_{\epsilon\to 0}\frac{1}{4\pi\epsilon}\int_{-\infty}^{\infty}(2+we^{-iuy}+\overline{w}e^{iuy})|s(u+\epsilon)-s(u-\epsilon)|^2\,du.$$

If we take w successively to equal ± 1, $\pm i$, and combine four expressions of this sort as in (21·09), we obtain:

THEOREM 27. *If $f(x)$ belongs to S,*

(21·17) $$\phi(y) = \lim_{\epsilon\to 0}\frac{1}{4\pi\epsilon}\int_{-\infty}^{\infty} e^{iuy}|s(u+\epsilon)-s(u-\epsilon)|^2\,du.$$

GENERALIZED HARMONIC ANALYSIS

Let us now consider the function

(21·175) $\quad \phi_\epsilon(y) = \dfrac{1}{4\pi\epsilon} \displaystyle\int_{-\infty}^{\infty} e^{iuy} |s(u+\epsilon) - s(u-\epsilon)|^2 du.$

It follows at once from the definition that

$$|\phi_\epsilon(y)| \leq \phi_\epsilon(0).$$

Inasmuch as $\phi_\epsilon(0)$ tends to $\phi(0)$ as $\epsilon \to 0$, $\phi_\epsilon(y)$ is bounded for all values of y and small values of ϵ. It thus tends *boundedly* to $\phi(y)$ as $\epsilon \to 0$ if $f(x)$ belongs to S. Thus, by bounded convergence,

$$\frac{1}{2\eta}\int_{-\eta}^{\eta} \phi(y)\, dy = \lim_{\epsilon \to 0} \frac{1}{2\eta}\int_{-\eta}^{\eta} \phi_\epsilon(y)\, dy$$

$$= \lim_{\epsilon \to 0} \frac{1}{2\eta}\int_{-\eta}^{\eta} dy \, \frac{1}{4\pi\epsilon}$$

$$\times \int_{-\infty}^{\infty} e^{iuy} |s(u+\epsilon) - s(u-\epsilon)|^2 du.$$

As this repeated integral converges absolutely, we may invert the order of integration, and we get

$$\frac{1}{2\eta}\int_{-\eta}^{\eta} \phi(y)\, dy$$

$$= \lim_{\epsilon \to 0} \frac{1}{4\pi\epsilon}\int_{-\infty}^{\infty} |s(u+\epsilon) - s(u-\epsilon)|^2 du \, \frac{1}{2\eta}\int_{-\eta}^{\eta} e^{iuy}\, dy$$

$$= \lim_{\epsilon \to 0} \frac{1}{4\pi\epsilon}\int_{-\infty}^{\infty} \frac{\sin u\eta}{u\eta} |s(u+\epsilon) - s(u-\epsilon)|^2 du.$$

Similarly,

$$\frac{1}{\eta}\int_{-\eta}^{\eta}\left(1 - \frac{|y|}{\eta}\right)\phi(y)\, dy$$

$$= \lim_{\epsilon \to 0} \frac{1}{\eta}\int_{-\eta}^{\eta}\left(1 - \frac{|y|}{\eta}\right)\phi_\epsilon(y)\, dy$$

$$= \lim_{\epsilon \to 0} \frac{1}{\eta}\int_{-\eta}^{\eta}\left(1 - \frac{|y|}{\eta}\right) dy \, \frac{1}{4\pi\epsilon}\int_{-\infty}^{\infty} e^{iuy} |s(u+\epsilon) - s(u-\epsilon)|^2 du$$

$$= \lim_{\epsilon \to 0} \frac{1}{4\pi\epsilon}\int_{-\infty}^{\infty} |s(u+\epsilon) - s(u-\epsilon)|^2 du \, \frac{1}{\eta}\int_{-\eta}^{\eta}\left(1 - \frac{|y|}{\eta}\right) e^{-iuy}\, dy$$

$$= \lim_{\epsilon \to 0} \frac{1}{4\pi\epsilon}\int_{-\infty}^{\infty} \frac{4\sin^2 \frac{u\eta}{2}}{u^2 \eta^2} |s(u+\epsilon) - s(u-\epsilon)|^2 du.$$

GENERALIZED HARMONIC ANALYSIS

Thus, if $f(x)$ belongs to S', we have

$$\phi(0) = \lim_{\eta \to 0} \frac{1}{\eta} \int_{-\eta}^{\eta} \left(1 - \frac{|y|}{\eta}\right) \phi(y)\, dy,$$

and

(21·18)
$$\lim_{\eta \to 0} \lim_{\epsilon \to 0} \frac{1}{4\pi\epsilon} \int_{-\infty}^{\infty} \left[1 - \frac{4\sin^2 \frac{u\eta}{2}}{u^2 \eta^2}\right] |s(u+\epsilon) - s(u-\epsilon)|^2\, du = 0.$$

Since, when $|u\eta| > \pi$,

$$1 - \frac{4\sin^2 \frac{u\eta}{2}}{u^2 \eta^2} > 1 - \frac{4}{\pi^2},$$

it follows from the positiveness of the integrand in (21·18) that

$$\varlimsup_{\eta \to 0} \varlimsup_{\epsilon \to 0} \frac{1}{4\pi\epsilon} \left[\int_{\pi/\eta}^{\infty} + \int_{-\infty}^{-\pi/\eta}\right] |s(u+\epsilon) - s(u-\epsilon)|^2\, du = 0,$$

or simply

(21·19)
$$\lim_{A \to \infty} \varlimsup_{\epsilon \to 0} \frac{1}{4\pi\epsilon} \left[\int_{A}^{\infty} + \int_{-\infty}^{-A}\right] |s(u+\epsilon) - s(u-\epsilon)|^2\, du = 0.$$

Again, let $f(x)$ belong to S and let (21·19) hold. By (21·17)

(21·20) $|\phi(y) - \phi(0)|$
$$\leq 2 \varlimsup_{\epsilon \to 0} \left[\int_{A}^{\infty} + \int_{-\infty}^{-A}\right] |s(u+\epsilon) - s(u-\epsilon)|^2\, du$$
$$+ \varlimsup_{\epsilon \to 0} \int_{-A}^{A} (e^{iuy} - 1) |s(u+\epsilon) - s(u-\epsilon)|^2\, du.$$

Let us choose A, as is possible by (21·19), so large that the first term on the right-hand side of (21·20) does not exceed $\delta/2$. Let us then choose $|y|$ so small that over $(-A, A)$

$$|e^{-iuy} - 1| \leq \delta/(2\phi(0)).$$

Then we shall have $\quad |\phi(y) - \phi(0)| \leq \delta$.

We thus have proved:

THEOREM 28. *If $f(x)$ belongs to S, it will belong to S' when and only when* (21·19) *is true.*

GENERALIZED HARMONIC ANALYSIS

We now introduce two new definitions:

$$(21\cdot 21) \quad \sigma(u) = \frac{1}{\sqrt{2\pi}} \left\{ \underset{A \to \infty}{\text{l.i.m.}} \left[\int_1^A + \int_{-A}^{-1} \right] \frac{\phi(\xi) e^{-iu\xi}}{-i\xi} d\xi \right.$$
$$\left. + \int_{-1}^1 \phi(\xi) \frac{e^{-iu\xi} - 1}{-i\xi} d\xi \right\},$$

and

$$(21\cdot 22) \quad \sigma_\epsilon(u) = \frac{1}{\sqrt{2\pi}} \left\{ \underset{A \to \infty}{\text{l.i.m.}} \left[\int_1^A + \int_{-A}^{-1} \right] \frac{\phi_\epsilon(\xi) e^{-iu\xi}}{-i\xi} d\xi \right.$$
$$\left. + \int_{-1}^1 \phi_\epsilon(\xi) \frac{e^{-iu\xi} - 1}{-i\xi} d\xi \right\}.$$

Both these expressions will exist, as $\phi(\xi)$ and $\phi_\epsilon(\xi)$ are bounded. Moreover, by (21·175) and the Plancherel theorem,

$$(21\cdot 23) \quad \underset{A \to \infty}{\text{l.i.m.}} \frac{1}{\sqrt{2\pi}} \int_{-A}^A \phi_\epsilon(\xi) e^{-iu\xi} d\xi$$
$$= \frac{1}{2\epsilon \sqrt{2\pi}} |s(u+\epsilon) - s(u-\epsilon)|^2,$$

since the latter function belongs to L_2, being the square of a bounded function belonging to L_2, and thus being dominated by a constant times that function. By a simple application of X_{41a}, if we integrate both sides, we get

$$(21\cdot 24) \quad \frac{1}{\sqrt{2\pi}} \int_{-\infty}^\infty \phi_\epsilon(\xi) \frac{e^{-iu\xi} - 1}{-i\xi} d\xi$$
$$= \int_0^u \frac{1}{2\epsilon \sqrt{2\pi}} |s(u+\epsilon) - s(u-\epsilon)|^2 du.$$

If we make use of the principle (X_{41}) that when a function approaches one limit in the ordinary sense and another limit in the mean, these limits are almost everywhere the same, (21·22) and (21·24) yield us

$$(21\cdot 25) \quad \sigma_\epsilon(u) = \text{const.} + \int_0^u \frac{1}{2\epsilon \sqrt{2\pi}} |s(u+\epsilon) - s(u-\epsilon)|^2 du.$$

Since $\phi_\epsilon(\xi)$ tends boundedly to $\phi(\xi)$ as $\epsilon \to 0$, $\phi_\epsilon(\xi)/(-i\xi)$ tends in the mean to $\phi(\xi)/(-i\xi)$ over any range not containing the origin. From these facts we may readily conclude that over any finite range of u,

(21·255) $$\sigma(u) = \underset{\epsilon \to 0}{\text{l.i.m.}}\ \sigma_\epsilon(u).$$

There is no trouble in concluding directly from bounded convergence that the second integral of (21·22) tends uniformly in u to that in (21·21) as $\epsilon \to 0$, while as to the first integrals in the two formulas, our argument depends on the fact that convergence in the mean is invariant under a Fourier transformation.

It follows that

(21·257)
$$\sigma(u) = \text{const.} + \underset{\epsilon \to 0}{\text{l.i.m.}}\ \frac{1}{2\epsilon\sqrt{2\pi}} \int_0^u |s(u+\epsilon) - s(u-\epsilon)|^2 du.$$

The constant in this formula may readily be verified to be

$$-\left[\int_1^A + \int_{-A}^{-1}\right] \frac{\phi(\xi)}{i\xi} d\xi.$$

It will be seen that

(21·26) $\sigma(u) - \sigma(-u)$
$$= \underset{\epsilon \to 0}{\text{l.i.m.}}\ \frac{1}{2\epsilon\sqrt{2\pi}} \int_{-u}^u |s(u+\epsilon) - s(u-\epsilon)|^2 du.$$

In the case where $f(x)$ is taken to be $\sum_1^n A_j{}^{-u} e^{i\Lambda_j x}$, we have

$$\sigma(u) = \sum_1^n |A_j|^2 \tau_j(u),$$

where $\tau_j(u) = \dfrac{1}{\sqrt{2\pi}} \left\{ \underset{A \to \infty}{\text{l.i.m.}} \left[\int_1^A + \int_{-A}^{-1} \right] \dfrac{e^{i(\Lambda_j - u)\xi}}{-i\xi} d\xi \right.$

$$\left. + \int_{-1}^1 \frac{e^{i(\Lambda_j - u)\xi} - e^{i\Lambda_j \xi}}{-i\xi} d\xi \right\}$$

$$= \text{const.} + \frac{1}{\sqrt{2\pi}} \int_{-\infty}^\infty \frac{e^{i(\Lambda_j - u)\xi} - e^{i\Lambda_j \xi}}{-i\xi} d\xi.$$

Now $\displaystyle\int_{-\infty}^\infty \frac{e^{iv\xi} - 1}{i\xi} d\xi = \int_{-\infty}^\infty \frac{\sin v\xi}{\xi} d\xi = \pi\,\text{sgn}\,v.$

Hence $\tau_j(u) = \text{const.} + \sqrt{\dfrac{\pi}{2}}\,(\text{sgn}\,\Lambda_j - \text{sgn}(\Lambda_j - u)),$

and this is a step-function making an upward jump of $\sqrt{2\pi}$ at $u = \Lambda_j$. Thus $\sigma(u)$ is a step-function making a jump of $\sqrt{2\pi}\,|A_j|^2$ at every point $u = \Lambda_j$ and constant elsewhere. The

GENERALIZED HARMONIC ANALYSIS

total increment of $\sigma(u)$ over any interval corresponds to $\sqrt{2\pi}$ multiplied by the sum of the squares of the amplitudes of the trigonometric terms of $f(x)$ with frequencies lying within that interval. Physically speaking, this is the total energy of that portion of the oscillation $x = f(t)$ lying within the interval in question, x being taken to be a displacement and t the time. As $\sigma(u)$ determines the energy-distribution in the spectrum of $f(x)$, we may call it briefly the "spectrum" of $f(t)$.

The author sees no compelling reason to avoid a physical terminology in pure mathematics when a mathematical concept corresponds closely to a concept already familiar in physics. When a new term is to be invented to describe an idea new to the pure mathematician, it is by all means better to avoid needless duplication, and to choose the designation already current. The "spectrum" of this book merely amounts to rendering precise the notion familiar to the physicist, and may as well be known by the same name.

In the various cases of functions f and ϕ which we have considered in this section, $\sigma(u)$ is easy to determine. In case (i), $\sigma(u)$ is constant. In case (ii), $\sigma(u)$ makes a jump of $\sqrt{2\pi}$ at 0 and is otherwise constant. In case (iii), $\sigma(u)$ is constant. In case (iv), the Fourier transform of $\phi(x)$ is

$$\frac{1}{\sqrt{2\pi}} \int_{-1}^{1} (1 - |x|) e^{iux} dx = \frac{2(1 - \cos u)}{u^2},$$

and by an argument like that of (21·23), (21·24) and (21·257), we see that

$$\sigma(u) = \text{const.} + \int_{0}^{u} \frac{2(1 - \cos v)}{v^2} dv.$$

In this case $\sigma(u)$ is a differentiable monotone function, which means in the language of physics that the spectrum of $f(t)$ is continuous and possesses a spectrum density.

An example of a function with a continuous spectrum, but without a spectrum density at every point, has been given by Mahler*. Whether this type of spectrum has any physical importance is not yet known.

* Mahler 1.

§ 22. The Spectra of Certain Linear Transforms of a Function.

We now come to a new group of lemmas and theorems concerning the spectra of functions of the form

$$\int_{-\infty}^{\infty} K(x-\xi) f(\xi) \, d\xi,$$

where

(22·01) $$\frac{1}{2B} \int_{-B}^{B} |f(x)|^2 \, dx$$

is uniformly bounded in B. Our first lemma reads:

Lemma 29$_1$. *Let $f(x, u, \lambda)$ belong to L_2 as a function of u for every x and λ, and let it be measurable in (x, u). Let*

(22·02) $$\int_{-\infty}^{\infty} |f(x, u, \lambda)|^2 \, du$$

be bounded in x and λ, and let

(22·03) $$\int_{-\infty}^{\infty} |f(x, u, \lambda) - f(x, u)|^2 \, du \to 0$$

uniformly over any finite range of x as $\lambda \to \infty$. Let $K(x)$ belong to L_1. Let

$$\int_{-\infty}^{\infty} K(x) f(x, u, \lambda) \, dx$$

exist for all u and λ. Then, as $\lambda \to \infty$,

(22·04) $$\int_{-\infty}^{\infty} \left| \int_{-\infty}^{\infty} K(x) f(x, u, \lambda) \, dx - \int_{-\infty}^{\infty} K(x) f(x, u) \, dx \right|^2 du \to 0.$$

Let us first note that $f(x, u)$ belongs to L_2, and that by (22·03), for all x,

(22·05)
$$\int_{-\infty}^{\infty} |f(x, u)|^2 \, du \leqslant \limsup \int_{-\infty}^{\infty} |f(x, u, \lambda)|^2 \, du.$$

GENERALIZED HARMONIC ANALYSIS

Again, by the Schwarz inequality,

(22·06)
$$\left[\int_A^B |K(x)|\,dx\right]^2 \limsup \int_{-\infty}^{\infty} |f(x,u,\lambda)-f(x,u)|^2\,du$$
$$\geq \int_A^B |K(x)|\,dx \int_A^B |K(y)|\,dy \int_{-\infty}^{\infty} du\,|f(x,u,\lambda)-f(x,u)|$$
$$\times |f(y,u,\lambda)-f(y,u)|$$
$$= \int_{-\infty}^{\infty} du \int_A^B |K(x)|\,dx \int_A^B |K(y)|\,dy\,|f(x,u,\lambda)-f(x,u)|$$
$$\times |f(y,u,\lambda)-f(y,u)|$$
$$= \int_{-\infty}^{\infty} du \left[\int_A^B |K(x)|\,|f(x,u,\lambda)-f(x,u)|\,dx\right]^2$$
$$\geq \int_{-\infty}^{\infty} du \left|\int_A^B K(x)[f(x,u,\lambda)-f(x,u)]\,dx\right|^2.$$

The formula (22·06) holds in the sense that if the hypotheses of lemma 29_1 are assumed, every term in the formula exists and satisfies the indicated inequalities.

Since (22·02) is bounded and (22·05) is valid,
$$\int_{-\infty}^{\infty} |f(x,u,\lambda)-f(x,u)|^2\,du$$
is bounded in x and λ. Thus, by (22·06),
$$\int_{-\infty}^{\infty} du \left|\int_A^B K(x)[f(x,u,\lambda)-f(x,u)]\,dx\right|^2 < \text{const.} \left[\int_A^B |K(x)|\,dx\right]^2.$$

Let it be noted that ∞ is a possible value of B, or $-\infty$ of A. We conclude that
$$\lim_{A\to\infty} \int_{-\infty}^{\infty} du \left|\int_A^{\infty} K(x)[f(x,u,\lambda)-f(x,u)]\,dx\right|^2 = 0$$
uniformly in λ, and similarly that
$$\lim_{B\to-\infty} \int_{-\infty}^{\infty} du \left|\int_{-\infty}^B K(x)[f(x,u,\lambda)-f(x,u)]\,dx\right|^2 = 0$$
uniformly in λ.

Again, it follows from (22·06) and (22·03) that for fixed finite A and B,
$$\lim_{\lambda\to\infty} \int_{-\infty}^{\infty} du \left|\int_B^A K(x)[f(x,u,\lambda)-f(x,u)]\,dx\right|^2 = 0.$$

Let A and $-B$ be so great that, for all λ,

(22·07) $\quad \int_{-\infty}^{\infty} du \left| \int_{A}^{\infty} K(x)[f(x, u, \lambda) - f(x, u)]\, dx \right|^2 < \epsilon/9;$

$\int_{-\infty}^{\infty} du \left| \int_{-\infty}^{B} K(x)[f(x, u, \lambda) - f(x, u)]\, dx \right|^2 < \epsilon/9.$

Then let λ be so large that

(22·08) $\quad \int_{-\infty}^{\infty} du \left| \int_{B}^{A} K(x)[f(x, u, \lambda) - f(x, u)]\, dx \right|^2 < \epsilon/9.$

We shall have (by the Minkowski inequality and (22·07-08))

$$\int_{-\infty}^{\infty} du \left| \int_{-\infty}^{\infty} K(x)[f(x, u, \lambda) - f(x, u)]\, dx \right|^2 < \epsilon.$$

From this (22·04) follows at once, and lemma 29_1 is proved.

Lemma 29_2. *Let $f(x)$ be a measurable function for which*

(22·01) $\quad \dfrac{1}{2B} \int_{-B}^{B} |f(x)|^2\, dx$

is uniformly bounded in B. Let $s(u)$ be defined as in (20·04-06). Let $K(x)$ be a bounded measurable function for which

(22·09) $\quad \int_{-\infty}^{\infty} |K(x)|^2 (1 + x^2)\, dx < \infty.$

Then (22·10)

$$t_1(u) = \underset{A \to \infty}{\text{l.i.m.}} \frac{1}{\sqrt{2\pi}} \left[\int_{1}^{A} + \int_{-A}^{-1} \right] \frac{e^{-iux}}{-ix}\, dx \int_{-\infty}^{\infty} K(x - \xi) f(\xi)\, d\xi$$

and

(22·11) $\quad t_2(u) = \dfrac{1}{\sqrt{2\pi}} \displaystyle\int_{-1}^{1} \dfrac{e^{-iux} - 1}{-ix}\, dx \int_{-\infty}^{\infty} K(x - \xi) f(\xi)\, d\xi$

will exist. Let

(22·12) $\quad\quad\quad t(u) = t_1(u) + t_2(u).$

Then

(22·13) $\quad t(u + \epsilon) - t(u - \epsilon)$

$\quad\quad\quad - \{s(u + \epsilon) - s(u - \epsilon)\} \displaystyle\int_{-\infty}^{\infty} K(\xi) e^{-iu\xi}\, d\xi$

$= \underset{A \to \infty}{\text{l.i.m.}} \dfrac{1}{\sqrt{2\pi}} \displaystyle\int_{-\infty}^{\infty} K(\xi)\, d\xi \int_{-A}^{A} 2f(x - \xi)$

$\quad\quad\quad \times \left[\dfrac{\sin \epsilon x}{x} - \dfrac{\sin \epsilon (x - \xi)}{x - \xi} \right] e^{-iux}\, dx.$

GENERALIZED HARMONIC ANALYSIS

To begin with, by (20·07),

$$\{s(u+\epsilon) - s(u-\epsilon)\} \int_{-\infty}^{\infty} K(\xi) e^{-iu\xi} d\xi$$
$$= \int_{-\infty}^{\infty} K(\xi) e^{-iu\xi} d\xi \underset{A \to \infty}{\text{l.i.m.}} \frac{1}{\sqrt{2\pi}} \int_{-A}^{A} f(x) \frac{2 \sin \epsilon x}{x} e^{-iux} dx.$$

It is obvious on inspection, however, that the Fourier transform of $f(x - \xi)$ is $e^{iu\xi}$ times the Fourier transform of $f(x)$. Thus

(22·14) $\{s(u+\epsilon) - s(u-\epsilon)\} \int_{-\infty}^{\infty} K(\xi) e^{-iu\xi} d\xi$
$$= \int_{-\infty}^{\infty} K(\xi) d\xi \underset{A \to \infty}{\text{l.i.m.}} \frac{1}{\sqrt{2\pi}} \int_{-A}^{A} f(x - \xi) \frac{2 \sin \epsilon (x - \xi)}{x - \xi} e^{-iux} dx.$$

The function

(22·15) $$F(\xi, u, A) = \frac{1}{\sqrt{2\pi}} \int_{-A}^{A} f(x - \xi) \frac{2 \sin \epsilon (x - \xi)}{x - \xi} e^{-iux} dx$$

belongs to L_2 in u, for every ξ and A, and is thus measurable. We have

$$\int_{-\infty}^{\infty} |F(\xi, u, A)|^2 du \leq \int_{-\infty}^{\infty} |f(x)|^2 \frac{4 \sin^2 \epsilon x}{x^2} dx < \infty,$$

and $\underset{A \to \infty}{\text{l.i.m.}} F(\xi, u, A) = F(\xi, u)$ exists for every ξ. We have

$$\int_{-\infty}^{\infty} |F(\xi, u, A) - F(\xi, u)|^2 du$$
$$\leq \left[\int_{A-\xi}^{\infty} + \int_{-\infty}^{-A-\xi} \right] |f(\xi)|^2 \frac{4 \sin^2 \epsilon x}{x^2} dx,$$

and this tends uniformly to 0 over any finite range of ξ as $A \to \infty$. The functions $K(\xi)$ and $F(\xi, u, A)$ belong to L_2, and, by the Schwarz inequality,

$$\int_{-\infty}^{\infty} K(\xi) e^{-iu\xi} F(\xi, u, A) d\xi$$

exists for every u and A. An application of the Schwarz inequality to (22·09) will show that $K(\xi)$ belongs to L_1. Thus, by lemma 29_1,

$$\int_{-\infty}^{\infty} K(\xi) e^{-iu\xi} d\xi \underset{A \to \infty}{\text{l.i.m.}} F(\xi, u, A)$$
$$= \underset{A \to \infty}{\text{l.i.m.}} \int_{-\infty}^{\infty} K(\xi) e^{-iu\xi} F(\xi, u, A) d\xi.$$

Combining this with (22·14) and (22·15), we get

$$(22\cdot16) \quad \{s(u+\epsilon)-s(u-\epsilon)\}\int_{-\infty}^{\infty} K(\xi)\,e^{-iu\xi}\,d\xi$$
$$= \underset{A\to\infty}{\text{l.i.m.}} \int_{-\infty}^{\infty} K(\xi)\,d\xi \frac{1}{\sqrt{2\pi}} \int_{-A}^{A} f(x-\xi)\frac{2\sin\epsilon(x-\xi)}{x-\xi} e^{-iux}\,dx.$$

By the Schwarz inequality,

$$(22\cdot17) \quad \left[\int_{-\infty}^{\infty} |K(x-\xi)f(\xi)|\,d\xi\right]^2 \le \int_{-\infty}^{\infty} \frac{|f(\xi)|^2}{1+(x-\xi)^2}\,d\xi$$
$$\times \int_{-\infty}^{\infty} |K(x-\xi)|^2[1+(x-\xi)^2]\,d\xi.$$

By (22·09) this does not exceed an expression of the form

$$(22\cdot18) \quad \text{const.} \int_{-\infty}^{\infty} \frac{|f(\xi)|^2}{1+(x-\xi)^2}\,d\xi$$
$$= \text{const.} \int_{-\infty}^{\infty} \frac{1+\xi^2}{1+(x-\xi)^2} \frac{|f(\xi)|^2}{1+\xi^2}\,d\xi.$$

An elementary calculation will show that for all real ξ,

$$\frac{1+\xi^2}{1+(x-\xi)^2} \le \frac{\sqrt{x^2+4}+x}{\sqrt{x^2+4}-x} = \left(\frac{\sqrt{x^2+4}+x}{2}\right)^2.$$

Thus, by (22·17) and (20·03),

$$\int_{-\infty}^{\infty} |K(x-\xi)f(\xi)|\,d\xi \le \text{const.}\,x + \text{const.}$$

The double integral

$$(22\cdot19) \quad \int_{-A}^{A} \frac{2\sin\epsilon x}{x} e^{-iux}\,dx \int_{-\infty}^{\infty} K(\xi)f(x-\xi)\,d\xi$$

is thus absolutely convergent, and may be written

$$(22\cdot20) \quad \int_{-\infty}^{\infty} K(\xi)\,d\xi \int_{-A}^{A} f(x-\xi)\frac{2\sin\epsilon x}{x} e^{-iux}\,dx.$$

Again, by (22·18),

$$\frac{1}{2T}\int_{-T}^{T} dx \left|\int_{-\infty}^{\infty} K(x-\xi)f(\xi)\,d\xi\right|^2$$
$$\le \frac{\text{const.}}{2T} \int_{-T}^{T} dx \int_{-\infty}^{\infty} \frac{|f(\xi)|^2}{1+(x-\xi)^2}\,d\xi$$

GENERALIZED HARMONIC ANALYSIS

$$= \frac{\text{const.}}{2T} \int_{-\infty}^{\infty} |f(\xi)|^2 d\xi \int_{-T}^{T} \frac{dx}{1+(x-\xi)^2}$$

$$= \frac{\text{const.}}{2T} \int_{-\infty}^{\infty} |f(\xi)|^2 [\tan^{-1}(T+\xi) - \tan^{-1}(\xi-T)] d\xi$$

$$= \text{const.} \int_{-\infty}^{\infty} |f(\xi)|^2 \frac{1}{2T} \tan^{-1} \frac{2T}{1-T^2+\xi^2} d\xi.$$

In this last line, the $\tan^{-1}$ is taken to lie between 0 and π. Over the range of $\xi\,(-2T, 2T)$,

$$\frac{1}{2T} \tan^{-1} \frac{2T}{1-T^2+\xi^2} \leqslant \frac{\pi}{2T},$$

and outside this range,

$$\frac{1}{2T} \tan^{-1} \frac{2T}{1-T^2+\xi^2} \leqslant \frac{1}{2T} \tan^{-1} \frac{8T}{1+3\xi^2} \leqslant \frac{4}{1+3\xi^2} \leqslant \frac{4}{1+\xi^2}.$$

Thus

$$\frac{1}{2T} \int_{-T}^{T} dx \left| \int_{-\infty}^{\infty} K(y-\xi)f(\xi) d\xi \right|^2 \leqslant \text{const.} \left\{ \frac{\pi}{2T} \int_{-2T}^{2T} |f(\xi)|^2 d\xi \right.$$
$$\left. + 4 \int_{-\infty}^{\infty} \frac{|f(\xi)|^2}{1+\xi^2} d\xi \right\}.$$

If we apply (20·03) to the second term, we see that

$$(22\cdot21) \quad \frac{1}{2T} \int_{-T}^{T} dy \left| \int_{-\infty}^{\infty} K(x-\xi)f(\xi) d\xi \right|^2$$
$$\leqslant \text{const.} \limsup \frac{1}{2U} \int_{-U}^{U} |f(\xi)|^2 d\xi.$$

It is hence legitimate to treat $\int_{-\infty}^{\infty} K(x-\xi)f(\xi) d\xi$ exactly as we have treated $f(x)$ in §20, and the functions defined in (22·10–12) will exist exactly as do those in (20·04–06).

As in (20·07),

$$t(u+\epsilon) - t(u-\epsilon) = \underset{A \to \infty}{\text{l.i.m.}} \frac{1}{\sqrt{2\pi}} \int_{-A}^{A} \frac{2 \sin \epsilon x}{x} e^{-iux} dx$$
$$\times \int_{-\infty}^{\infty} K(\xi)f(x-\xi) d\xi,$$

and, as (22·19) and (22·20) are equivalent,

$$t(u+\epsilon) - t(u-\epsilon) = \underset{A \to \infty}{\text{l.i.m.}} \int_{-\infty}^{\infty} K(\xi) d\xi \int_{-A}^{A} f(x-\xi) \frac{2 \sin \epsilon x}{x} e^{-iux} dx.$$

170 GENERALIZED HARMONIC ANALYSIS

If we combine this with (22·16), we obtain (22·13). We have thus established lemma 29_2.

Lemma 29_3. *On the hypothesis of lemma 29_2, if $\xi K(\xi)$ belongs to L_1,*

$$(22\cdot22) \quad \lim_{\epsilon \to 0} \frac{1}{\epsilon} \int_{-\infty}^{\infty} \left| t(u+\epsilon) - t(u-\epsilon) - \{s(u+\epsilon) - s(u-\epsilon)\} \right. $$
$$\left. \times \int_{-\infty}^{\infty} K(\xi) e^{-iu\xi} d\xi \right|^2 du = 0.$$

To begin with, by (21·14) and (21·15),

$$(22\cdot23) \quad \left[\int_A^\infty + \int_{-\infty}^{-A}\right] \left| 2f(x-\xi) \left[\frac{\sin \epsilon x}{x} - \frac{\sin \epsilon (x-\xi)}{x-\xi}\right]\right|^2 dx$$
$$\leqslant \left[\int_A^\infty + \int_{-\infty}^{-A}\right] \left|\frac{16\epsilon\xi}{|x|+|\xi|} f(x-\xi)\right|^2 dx$$
$$= \left[\int_{A-\xi}^\infty + \int_{-\infty}^{-A-\xi}\right] \left|\frac{16\epsilon\xi}{|x+\xi|+|\xi|} f(x)\right|^2 dx$$
$$\leqslant \left[\int_{A-|\xi|}^\infty + \int_{-\infty}^{-A+|\xi|}\right] 256\epsilon^2 \xi^2 \frac{|f(x)|^2}{x^2} dx.$$

If we make a Fourier transformation, and remember that this leaves the integral of the square of the modulus of a function invariant, we see that

$$(22\cdot24) \quad \int_{-\infty}^{\infty} \left| \frac{1}{\epsilon\xi\sqrt{2\pi}} \operatorname*{l.i.m.}_{B \to \infty} \int_{-B}^{B} 2f(x-\xi) \right.$$
$$\times \left[\frac{\sin \epsilon x}{x} - \frac{\sin \epsilon(x-\xi)}{x-\xi}\right] e^{-iux} dx$$
$$\left. - \frac{1}{\epsilon\xi\sqrt{2\pi}} \int_{-A}^{A} 2f(x-\xi) \left[\frac{\sin \epsilon x}{x} - \frac{\sin \epsilon(x-\xi)}{x-\xi}\right] e^{-iux} dx \right|^2 du$$
$$\leqslant \left[\int_{A-|\xi|}^\infty + \int_{-\infty}^{-A+|\xi|}\right] \frac{256\,|f(x)|^2}{x^2} dx,$$

which tends to 0 uniformly for all ϵ and any finite range of ξ as $A \to \infty$. [Cf. (20·03).]

If we now take ξ to be the x of lemma 29_1, u to be the u, A to be the λ,

$$f(\xi, u, A) = \frac{1}{\sqrt{2\pi}} \int_{-A}^{A} 2f(x-\xi) \left[\frac{\sin \epsilon x}{x} - \frac{\sin \epsilon(x-\xi)}{x-\xi}\right] e^{-iux} dx,$$

GENERALIZED HARMONIC ANALYSIS

and $$f(\xi, u) = \underset{A \to \infty}{\text{l.i.m.}} f(\xi, u, A),$$

we shall find that the hypotheses of lemma 29_1 are satisfied—we have just established (22·03)—and that the conclusion, namely

$$\underset{A \to \infty}{\text{l.i.m.}} \frac{1}{\sqrt{2\pi}} \int_{-\infty}^{\infty} K(\xi) d\xi \int_{-A}^{A} 2f(x-\xi)$$
$$\times \left[\frac{\sin \epsilon x}{x} - \frac{\sin \epsilon (x-\xi)}{x-\xi} \right] e^{-iux} dx$$
$$= \frac{1}{\sqrt{2\pi}} \int_{-\infty}^{\infty} K(\xi) d\xi \underset{A \to \infty}{\text{l.i.m.}} \int_{-A}^{A} 2f(x-\xi)$$
$$\times \left[\frac{\sin \epsilon x}{x} - \frac{\sin \epsilon (x-\xi)}{x-\xi} \right] e^{-iux} dx,$$

is valid. Thus, by (22·13),

(22·25)
$$t(u+\epsilon) - t(u-\epsilon) - \{s(u+\epsilon) - s(u-\epsilon)\} \int_{-\infty}^{\infty} K(\xi) e^{-iu\xi} d\xi$$
$$= \frac{1}{\sqrt{2\pi}} \int_{-\infty}^{\infty} K(\xi) d\xi \underset{A \to \infty}{\text{l.i.m.}} \int_{-A}^{A} 2f(x-\xi)$$
$$\times \left[\frac{\sin \epsilon x}{x} - \frac{\sin \epsilon (x-\xi)}{x-\xi} \right] e^{-iux} dx.$$

It follows from (21·14) and (21·15) that

(22·26) $$\frac{1}{\epsilon \xi \sqrt{2\pi}} \int_{-A}^{A} 2f(x-\xi) \left[\frac{\sin \epsilon x}{x} - \frac{\sin \epsilon (x-\xi)}{x-\xi} \right] e^{-iux} dx$$

is the Fourier transform of a function of L_2, and hence belongs to L_2. By (22·24), the same is true of

(22·27)
$$\underset{A \to \infty}{\text{l.i.m.}} \frac{1}{\epsilon \xi \sqrt{2\pi}} \int_{-A}^{A} 2f(x-\xi) \left[\frac{\sin \epsilon x}{x} - \frac{\sin \epsilon (x-\xi)}{x-\xi} \right] e^{-iux} dx.$$

It also is easy to show that over any finite range of ξ, the integral with respect to u of the square of the modulus of (22·26) is uniformly bounded in ϵ and ξ. A reference to (22·23) will show this. By (22·24), this is also true of (22·26). Thus

(22·26) $$\underset{\epsilon \to 0}{\text{l.i.m.}} \frac{1}{\epsilon^{\frac{1}{2}} \xi} \underset{A \to \infty}{\text{l.i.m.}} \frac{1}{\sqrt{2\pi}} \int_{-A}^{A} 2f(x-\xi)$$
$$\times \left[\frac{\sin \epsilon x}{x} - \frac{\sin \epsilon (x-\xi)}{x-\xi} \right] e^{-iux} dx = 0,$$

and this holds uniformly in ξ over any finite range. A further application of lemma 29_1, in which we now take ϵ in place of λ, and $\xi K(\xi)$ in place of $K(\xi)$, will show that

$$(22\cdot 28) \quad \underset{\epsilon \to 0}{\text{l.i.m.}} \int_{-\infty}^{\infty} K(\xi)\, d\xi \underset{A \to \infty}{\text{l.i.m.}} \frac{1}{\epsilon^{\frac{1}{2}}\sqrt{2\pi}} \int_{-A}^{A} 2f(x-\xi)$$
$$\times \left[\frac{\sin \epsilon x}{x} - \frac{\sin \epsilon(x-\xi)}{x-\xi}\right] e^{-iux}\, dx = 0,$$

and, by (22·25),

$$\underset{\epsilon \to 0}{\text{l.i.m.}} \frac{1}{\epsilon^{\frac{1}{2}}} \Big\{ t(u+\epsilon) - t(u-\epsilon) - \{s(u+\epsilon) - s(u-\epsilon)\}$$
$$\times \int_{-\infty}^{\infty} K(\xi)\, e^{-iu\xi}\, d\xi \Big\} = 0.$$

This again is merely a way of writing

$$(22\cdot 22) \quad \lim_{\epsilon \to 0} \frac{1}{\epsilon} \int_{-\infty}^{\infty} \Big| t(u+\epsilon) - t(u-\epsilon) - \{s(u+\epsilon) - s(u-\epsilon)\}$$
$$\times \int_{-\infty}^{\infty} K(\xi)\, e^{-iu\xi}\, d\xi \Big|^2 du = 0.$$

Lemma 29₄. *Let $f(x)$ be a measurable function for which*

$$(22\cdot 01) \quad \frac{1}{2B} \int_{-B}^{B} |f(x)|^2\, dx$$

is uniformly bounded in B. Let $s(u)$ be defined as in (20·04–06). Let $xK(x)$ belong to L_1, and $(1+|x|)K(x)$ to L_2. Let

$$(22\cdot 29) \quad \lim_{A \to \infty} \frac{1}{A} \int_{-A}^{A} dx \left| \int_{-\infty}^{\infty} K(x-\xi) f(\xi)\, d\xi \right|^2 = 0.$$

Then

$$(22\cdot 30) \quad \lim_{\epsilon \to 0} \frac{1}{\epsilon} \int_{-\infty}^{\infty} |s(u+\epsilon) - s(u-\epsilon)|^2$$
$$\times \left| \int_{-\infty}^{\infty} K(\xi)\, e^{-iu\xi}\, d\xi \right|^2 du = 0.$$

This is obtained by applying the Minkowski inequality to (22·22) and the formula

$$\lim_{\epsilon \to 0} \frac{1}{\epsilon} \int_{-\infty}^{\infty} |t(u+\epsilon) - t(u-\epsilon)|^2\, du = 0.$$

This in turn arises from applying theorem 22 to $t(u)$ instead of $s(u)$ and making use of (22·29).

GENERALIZED HARMONIC ANALYSIS

Lemma 29₅. *On the hypothesis of lemma* 29_4, *if there is no real u for which*

(22·31) $$\int_{-\infty}^{\infty} K(\xi) e^{-iu\xi} d\xi = 0,$$

then, for any finite C,

(22·32) $$\lim_{\epsilon \to 0} \frac{1}{\epsilon} \int_{-C}^{C} |s(u+\epsilon) - s(u-\epsilon)|^2 du = 0.$$

For over any finite range, $\int_{-\infty}^{\infty} K(\xi) e^{-iu\xi} d\xi$ is continuous and must exceed some constant Q in modulus. Thus, by (22·30),

$$\overline{\lim_{\epsilon \to 0}} \frac{Q^2}{\epsilon} \int_{-\infty}^{\infty} |s(u+\epsilon) - s(u-\epsilon)|^2 du$$

$$\leq \lim_{\epsilon \to 0} \frac{1}{\epsilon} \int_{-\infty}^{\infty} |s(u+\epsilon) - s(u-\epsilon)|^2 \left| \int_{-\infty}^{\infty} K(\xi) e^{-iu\xi} d\xi \right|^2 du = 0.$$

Lemma 29₆. *Let $f(x)$ belong to S. Let $xK(x)$ belong to L_1, and $(1+|x|)K(x)$ to L_2. Let*

$$g(x) = \int_{-\infty}^{\infty} K(x-\xi) f(\xi) d\xi.$$

Then $g(x)$ belongs to S'.

Let $s(u)$ be defined as in (20·04–06), and let $t(u)$ be defined as in (22·10–12). Let

$$K_3(x) = \frac{1}{\eta} \int_0^{\eta} K(x+\xi) d\xi.$$

By the hypothesis of lemma 29_6, $xK_3(x)$ will belong to L_1 and $(1+|x|)K_3(x)$ to L_2, for

$$\int_{-\infty}^{\infty} \frac{|x|}{\eta} dx \left| \int_0^{\eta} K(x+\xi) d\xi \right| \leq \int_{-\infty}^{\infty} \frac{|x|}{\eta} dx \int_0^{\eta} |K(x+\xi)| d\xi$$

$$\leq \int_{-\infty}^{\infty} \frac{dx}{\eta} \int_0^{\eta} (|x+\xi| + \eta) |K(x+\xi)| d\xi$$

$$\leq \int_{-\infty}^{\infty} (1+|x|) |K(x)| dx,$$

and $\int_{-\infty}^{\infty} \frac{(1+|x|)^2}{\eta^2} dx \left| \int_0^{\eta} K(x+\xi) d\xi \right|^2$

$$\leq \int_0^{\eta} d\xi_1 \int_0^{\eta} d\xi_2 \int_{-\infty}^{\infty} \frac{(1+|x|)^2}{\eta^2} |K(x+\xi_1)| |K(x+\xi_2)| dx$$

$$\leqslant \max_{0<\lambda<\eta} \int_{-\infty}^{\infty} (1+|x|)^2 |K(x+\lambda)|^2 dx$$

$$\leqslant \int_{-\infty}^{\infty} (1+\eta+|x|)^2 |K(x)|^2 dx$$

$$\leqslant (1+\eta)^2 \int_{-\infty}^{\infty} (1+|x|)^2 |K(x)|^2 dx.$$

If $\mu(u)$ corresponds to K_3 as $t(u)$ to K, we then have, by lemma 29_3,

$$(22\cdot33) \quad \lim_{\epsilon \to 0} \frac{1}{\epsilon} \int_{-\infty}^{\infty} \left| \mu(u+\epsilon) - \mu(u-\epsilon) \right.$$

$$\left. - \{s(u+\epsilon) - s(u-\epsilon)\} \int_{-\infty}^{\infty} K_3(\xi) e^{-iu\xi} d\xi \right|^2 du = 0,$$

and since

$$\int_{-\infty}^{\infty} K_3(\xi) e^{-iu\xi} d\xi = \frac{1}{\eta} \int_0^{\eta} dx \int_{-\infty}^{\infty} K(\xi) e^{-iu(\xi-x)} d\xi$$

$$= \frac{e^{iu\eta} - 1}{iu\eta} \int_{-\infty}^{\infty} K(\xi) e^{-iu\xi} d\xi,$$

$(22\cdot33)$ becomes

$$(22\cdot34) \quad \lim_{\epsilon \to 0} \frac{1}{\epsilon} \int_{-\infty}^{\infty} \left| \mu(u+\epsilon) - \mu(u-\epsilon) \right.$$

$$\left. - \{s(u+\epsilon) - s(u-\epsilon)\} \frac{e^{iu\eta}-1}{iu\eta} \int_{-\infty}^{\infty} K(\xi) e^{-iu\xi} d\xi \right|^2 du = 0.$$

On the other hand, by lemma 29_3,

$$\lim_{\epsilon \to 0} \frac{1}{\epsilon} \int_{-\infty}^{\infty} \left| \tau(u+\epsilon) - \tau(u-\epsilon) \right.$$

$$\left. - \{s(u+\epsilon) - s(u-\epsilon)\} \int_{-\infty}^{\infty} K(\xi) e^{-iu\xi} d\xi \right|^2 = 0,$$

and hence

$$\lim_{\epsilon \to 0} \frac{1}{\epsilon} \int_{-\infty}^{\infty} \left| \frac{e^{iu\eta}-1}{iu\eta} \{\tau(u+\epsilon) - \tau(u-\epsilon)\} \right.$$

$$\left. - \{s(u+\epsilon) - s(u-\epsilon)\} \frac{e^{iu\eta}-1}{iu\eta} \int_{-\infty}^{\infty} K(\xi) e^{-iu\xi} d\xi \right|^2 = 0.$$

Combining this with $(21\cdot34)$ by means of the Minkowski inequality, we obtain

$$(22\cdot35) \quad \lim_{\epsilon \to 0} \frac{1}{\epsilon} \int_{-\infty}^{\infty} \left| \frac{e^{iu\eta}-1}{iu\eta} \{\tau(u+\epsilon) - \tau(u-\epsilon)\} \right.$$

$$\left. - \mu(u+\epsilon) + \mu(u-\epsilon) \right|^2 du = 0.$$

GENERALIZED HARMONIC ANALYSIS

Again, by (20·03), (22·17), (22·21), and the reasoning on pp. 168 and 169,

$$(22·36) \quad \frac{1}{2T}\int_{-T}^{T} dx \left| \int_{-\infty}^{\infty} [K(x-\xi) - K_3(x-\xi)] f(\xi) d\xi \right|^2$$

$$\leqslant \text{const. } \limsup \frac{1}{2U} \int_{-U}^{U} |f(\xi)|^2 d\xi$$

$$\times \int_{-\infty}^{\infty} |K(x-\xi) - K_3(x-\xi)|^2 [1 + (x-\xi)^2] d\xi$$

$$= \text{const.} \int_{-\infty}^{\infty} \left| K(x) - \frac{1}{\eta}\int_{0}^{\eta} K(x+\xi) d\xi \right|^2 (1+x^2) dx$$

$$= \frac{\text{const.}}{\eta^2} \int_{0}^{\eta} d\xi_1 \int_{0}^{\eta} d\xi_2 \int_{-\infty}^{\infty} |K(x) - K(x+\xi_1)|$$

$$\times |K(x) - K(x+\xi_2)| (1+x^2) dx$$

$$\leqslant \text{const. } \limsup_{0 \leqslant \xi \leqslant \eta} \int_{-\infty}^{\infty} |K(x) - K(x+\xi)|^2 (1+x^2) dx \, ^*$$

$$\leqslant \text{const. } \limsup_{0 \leqslant \xi \leqslant \eta} \left\{ \int_{-\infty}^{\infty} |K(x)(1+|x|) \right.$$

$$- K(x+\xi)(1+|x+\xi|)|^2 dx$$

$$\left. + \int_{-\infty}^{\infty} |K(x+\xi)(|x+\xi| - |x|)|^2 dx \right\}.$$

The first term of the last formula of (22·36) vanishes as $\eta \to 0$ by X_{31}, and the second term is dominated by $\xi^2 \int_{-\infty}^{\infty} |K(x)|^2 dx$, and likewise vanishes. Thus

$$\lim_{\eta \to 0} \lim_{T \to \infty} \frac{1}{2T}\int_{-T}^{T} dx \left| \int_{-\infty}^{\infty} [K(x-\xi) - K_3(x-\xi)] f(\xi) d\xi \right|^2 = 0,$$

or, by theorem 22,

$$\lim_{\eta \to 0} \lim_{\epsilon \to 0} \frac{1}{\epsilon} \int_{-\infty}^{\infty} |\tau(u+\epsilon) - \tau(u-\epsilon) - \mu(u+\epsilon) + \mu(u-\epsilon)|^2 du = 0.$$

If we combine this with (22·35) and make use of the Minkowski inequality, we obtain

$$\lim_{\eta \to 0} \lim_{\epsilon \to 0} \frac{1}{\epsilon} \int_{-\infty}^{\infty} \left| 1 - \frac{e^{iu\eta}-1}{iu\eta} \right|^2 |\tau(u+\epsilon) - \tau(u-\epsilon)|^2 du = 0.$$

* Here we employ the Schwarz inequality.

Now, if $|u\eta| > 4$,
$$\left|1 - \frac{e^{iu\eta}-1}{iu\eta}\right| \geq 1 - \tfrac{2}{4} = \tfrac{1}{2}.$$
Thus

(22·37)
$$\lim_{\eta \to 0} \overline{\lim_{\epsilon \to 0}} \frac{1}{2\epsilon} \left[\int_{4/\eta}^{\infty} + \int_{-\infty}^{-4/\eta}\right] |\tau(u+\epsilon) - \tau(u-\epsilon)|^2 \, du = 0.$$

By theorem 28, $g(x)$ will then belong to S' if it belongs to S.

It only remains to prove that $g(x)$ belongs to S. This will be the case if

$$\lim_{T \to \infty} \frac{1}{2T} \int_{-T}^{T} dx \int_{-\infty}^{\infty} K_1(x+\lambda-\xi) f(\xi) \, d\xi \int_{-\infty}^{\infty} \bar{K}_1(x-\xi) \bar{f}(\xi) \, dx$$

exists for every real λ. However,

(22·38)
$$\frac{1}{2T} \int_{-T}^{T} dx \int_{-\infty}^{\infty} K_1(x+\lambda-\xi) f(\xi) \, d\xi \int_{-\infty}^{\infty} \bar{K}_1(x-\xi) \bar{f}(\xi) \, d\xi$$
$$= \int_{-\infty}^{\infty} K_1(\xi) \, d\xi \int_{-\infty}^{\infty} \bar{K}_1(\eta) \, d\eta \, \frac{1}{2T} \int_{-T}^{T} f(x+\lambda-\xi) \bar{f}(x-\eta) \, dx.$$

The expression

(22·39)
$$\frac{1}{2T} \int_{-T}^{T} f(x+\lambda-\xi) \bar{f}(x-\eta) \, dx$$

converges to $\phi(x+\eta-\xi)$ (cf. lemma 26_1) and is dominated by

$$\frac{1}{2T} \sqrt{\int_{-T-|\lambda-\xi|}^{T+|\lambda-\xi|} |f(x)|^2 \, dx \int_{-T-|\eta|}^{T+|\eta|} |f(x)|^2 \, dx},$$

as one may see by applying the Schwarz inequality. This in turn is less than an expression of the form

$$\text{const.} \sqrt{\left(1 + \frac{|\lambda-\xi|}{T}\right)\left(1 + \frac{|\eta|}{T}\right)},$$

as results from the boundedness of $\frac{1}{2U} \int_{-U}^{U} |f(x)|^2 \, dx$. Thus (22·39) is dominated for all large enough values of T by

$$\text{const.} \sqrt{(1+|\xi|)(1+|\eta|)},$$

GENERALIZED HARMONIC ANALYSIS

and thus by const. $(1 + |\xi|)(1 + |\eta|)$.

Since $K(x)(1+|x|)$ belongs to L_1, the integrand (22·38) is dominated for all large values of T by a function with a finite integral, and thus, by dominated convergence,

$$\lim_{T \to \infty} \frac{1}{2T} \int_{-T}^{T} dx \int_{-\infty}^{\infty} K_1(x+\lambda-\xi) f(\xi) d\xi \int_{-\infty}^{\infty} \overline{K}_1(x-\xi) \overline{f}(\xi) d\xi$$
$$= \int_{-\infty}^{\infty} K_1(\xi) d\xi \int_{-\infty}^{\infty} K_1(\eta) \phi(\lambda+\eta-\xi) d\xi.$$

THEOREM 29. *Let $f(x)$ be a measurable function for which*

(22·01) $$\frac{1}{2B} \int_{-B}^{B} |f(x)|^2 dx$$

is uniformly bounded in B. Let $xK_1(x)$ and $xK_2(x)$ belong to L_1, and let $(1+|x|)K_1(x)$ and $(1+|x|)K_2(x)$ belong to L_2. Let there be no real u for which

(22·31) $$\int_{-\infty}^{\infty} K(\xi) e^{-iu\xi} d\xi = 0.$$

Let

(22·40) $$\lim_{A \to \infty} \frac{1}{2A} \int_{-A}^{A} dx \left| \int_{-\infty}^{\infty} K_1(x-\xi) f(\xi) d\xi \right|^2 = 0.$$

Then

(22·41) $$\lim_{A \to \infty} \frac{1}{2A} \int_{-A}^{A} dx \left| \int_{-\infty}^{\infty} K_2(x-\xi) f(\xi) d\xi \right|^2 = 0.$$

It will be seen that this theorem is Tauberian in a generalized sense. Formulae (22·40-41) indicate that

$$\int_{-\infty}^{\infty} K_{1,2}(x-\xi) f(\xi) d\xi$$

are asymptotically small *on the average*. Formula (22·01) asserts that $f(x)$ is bounded on the average. Thus if a function is bounded on the average, and a weighted moving average of it is asymptotically small on the average, and if the Fourier transform of the kernel of this moving average does not vanish, all moving averages with kernels of a very general type are asymptotically small on the average.

178 GENERALIZED HARMONIC ANALYSIS

We define $s(u)$ as in (20·04–06), and put

$$\tau_1(u) = \underset{A\to\infty}{\text{l.i.m.}} \frac{1}{\sqrt{2\pi}} \left[\int_1^A + \int_{-A}^{-1}\right] \frac{e^{-iux}}{-ix} dx \int_{-\infty}^{\infty} K_2(x-\xi) f(\xi) d\xi;$$

$$\tau_2(u) = \frac{1}{\sqrt{2\pi}} \int_{-1}^{1} \frac{e^{-iux}-1}{-ix} dx \int_{-\infty}^{\infty} K_2(x-\xi) f(\xi) d\xi;$$

$$\tau(u) = \tau_1(u) + \tau_2(u).$$

By (22·32),

$$\lim_{\epsilon \to 0} \frac{1}{\epsilon} \int_{-C}^{C} |s(u+\epsilon) - s(u-\epsilon)|^2 \left|\int_{-\infty}^{\infty} K_2(\xi) e^{-iu\xi}\right|^2 du = 0$$

for any positive C. Thus, by lemma 29₃ and the Minkowski inequality, we have respectively

$$\lim_{\epsilon\to 0} \frac{1}{\epsilon} \int_{-C}^{C} \left|\tau(u+\epsilon) - \tau(u-\epsilon) - \{s(u+\epsilon) - s(u-\epsilon)\}\right.$$
$$\left. \times \int_{-\infty}^{\infty} K_2(\xi) e^{-iu\xi} d\xi\right|^2 du = 0,$$

and

(22·42) $\quad \displaystyle\lim_{\epsilon\to 0} \frac{1}{\epsilon} \int_{-C}^{C} |\tau(u+\epsilon) - \tau(u-\epsilon)|^2 du = 0.$

The conditions under which we established (22·37) are satisfied. This gives us

$$\lim_{C\to\infty} \overline{\lim_{\epsilon\to 0}} \frac{1}{\epsilon} \left[\int_C^\infty + \int_{-\infty}^{-C}\right] |\tau(u+\epsilon) - \tau(u-\epsilon)|^2 du = 0.$$

If we combine this with (22·42), we get

$$\lim_{\epsilon\to 0} \frac{1}{\epsilon} \int_{-\infty}^{\infty} |\tau(u+\epsilon) - \tau(u-\epsilon)|^2 du = 0,$$

and hence, by theorem 22,

$$\lim_{A\to\infty} \frac{1}{2A} \int_{-A}^{A} dx \left|\int_{-\infty}^{\infty} K_2(x-\xi) f(\xi) d\xi\right|^2 = 0.$$

We have thus established theorem 29.

A closely allied theorem is:

THEOREM 30. *Let $f(x)$ belong to S. Let $xK(x)$ belong to L_1, and $(1+|x|)K(x)$ to L_2. Let*

$$g(x) = \int_{-\infty}^{\infty} K(x-\xi) f(\xi) d\xi.$$

GENERALIZED HARMONIC ANALYSIS

Let $\sigma(u)$ be defined as in (21·21), *and $\sigma_\epsilon(u)$ as in* (21·22). *Let $\psi(u)$ bear the same relation to g that $\sigma(u)$ does to f, and let $\psi_\epsilon(u)$ bear the same relation to g that $\sigma_\epsilon(u)$ does to f. Then*

$$(22\text{·}43) \quad \psi_\epsilon(u) = \text{const.} + \int_0^u \left| \int_{-\infty}^\infty K(\xi) e^{-iu\xi} d\xi \right|^2 d\sigma_\epsilon(u);$$

$$(22\text{·}44) \quad \psi(u) = \text{const.} + \int_0^u \left| \int_{-\infty}^\infty K(\xi) e^{-iu\xi} d\xi \right|^2 d\sigma(u).$$

For this we shall need the following lemma, which also serves as a basis for the theorems in the next section:

Lemma 30$_a$. *If $f(x)$ belongs to S, $\sigma(u)$ is real and may be so defined as to be monotone increasing.* Let us notice that as $\sigma(u)$ is indeterminate over an indeterminate null set, we cannot say that it will be monotone, but merely that it may so be defined.

By (21·25), $\sigma_\epsilon'(u)$ is real and positive or zero and $\sigma_\epsilon(u)$ is monotone increasing. By (21·255) and X$_{42}$, the result follows at once.

Let $s(u)$ be defined as in (20·04–06), and let $t(u)$ be defined as in (22·11–13). By (21·25),

$$\psi_\epsilon(u) = \text{const.} + \int_0^u \frac{1}{2\epsilon\sqrt{2\pi}} |t(u+\epsilon) - t(u-\epsilon)|^2 du.$$

If we combine this with (22·22), which holds by lemma 29$_3$, and invoke the aid of the Minkowski inequality, we have

$$(22\text{·}45) \quad \psi_\epsilon(u) = \text{const.} + \int_0^u \frac{1}{2\epsilon\sqrt{2\pi}} \Big| \{s(u+\epsilon) - s(u-\epsilon)\}$$
$$\times \int_{-\infty}^\infty K(\xi) e^{-iu\xi} \Big|^2 du.$$

By a further application of (21·25), we obtain

$$d\sigma_\epsilon(u) = \frac{1}{2\epsilon\sqrt{2\pi}} |s(u+\epsilon) - s(u-\epsilon)|^2 du,$$

which together with (22·45) yields us (22·43).

Clearly, as $xK(x)$ belongs to L_1,

$$\int_0^u du \int_{-\infty}^\infty xK(x) e^{-iux} dx = \int_{-\infty}^\infty xK(x) dx \int_0^u e^{-iux} du$$
$$= i \int_{-\infty}^\infty xK(x)(e^{-iux} - 1) dx.$$

From this it readily follows that $\int_{-\infty}^{\infty} K(x) e^{-iux} dx$ is a function with a bounded difference-quotient over any finite range of u, and hence that the same thing is true of $\left| \int_{-\infty}^{\infty} K(x) e^{-iux} dx \right|^2$. *A fortiori*, these functions are of limited total variation.

We may hence write (22·43) in the form

$$\psi_\epsilon(u) = \text{const.} + \sigma_\epsilon(u) \left| \int_{-\infty}^{\infty} K(\xi) e^{-iu\xi} d\xi \right|^2$$
$$- \int_0^u \sigma_\epsilon(u) \frac{d}{du} \left\{ \left| \int_{-\infty}^{\infty} K(\xi) e^{-iu\xi} d\xi \right|^2 \right\} du.$$

By (21·255), over any finite range of u,

$$\underset{\epsilon \to 0}{\text{l.i.m.}} \; \sigma_\epsilon(u) \left| \int_{-\infty}^{\infty} K(\xi) e^{-iu\xi} d\xi \right|^2 = \sigma(u) \left| \int_{-\infty}^{\infty} K(\xi) e^{-iu\xi} d\xi \right|^2.$$

By a second application of (21·255) in conjunction with X_{26},

$$\lim_{\epsilon \to 0} \int_0^u \sigma_\epsilon(u) \frac{d}{du} \left\{ \left| \int_{-\infty}^{\infty} K(\xi) e^{-iu\xi} d\xi \right|^2 \right\} du$$
$$= \int_0^u \sigma(u) \frac{d}{du} \left\{ \left| \int_{-\infty}^{\infty} K(\xi) e^{-iu\xi} d\xi \right|^2 \right\} du$$

uniformly over any finite range of u. By (21·255),

$$\psi(u) = \underset{\epsilon \to 0}{\text{l.i.m.}} \; \psi_\epsilon(u).$$

Combining these, almost everywhere,

$$\psi(u+\alpha) - \psi(u) = [\sigma(u+\alpha) - \sigma(u)] \left| \int_{-\infty}^{\infty} K(\xi) e^{-iu\xi} d\xi \right|^2$$
$$- \int_u^{u+\alpha} \sigma(u) \frac{d}{du} \left\{ \left| \int_{-\infty}^{\infty} K(\xi) e^{-iu\xi} d\xi \right|^2 \right\} du.$$

By lemma 30_a, we may integrate by parts, obtaining

$$\psi(u+\alpha) - \psi(u) = \int_u^{u+\alpha} \left| \int_{-\infty}^{\infty} K(\xi) e^{-iu\xi} d\xi \right|^2 d\sigma(u),$$

which is equivalent to (22·44).

§ 23. The Monotoneness of the Spectrum.

We now come to a group of theorems clustering around lemma 30_a, and making assertions concerning the function $\sigma(u)$.

GENERALIZED HARMONIC ANALYSIS

Theorem 31. *If $f(x)$ belongs to S, and $\sigma(u)$ is defined as in lemma 30_a,*

(23·01) $\qquad \sigma(\infty) - \sigma(-\infty) \leqslant \sqrt{2\pi}\,\phi(0).$

We shall have

(23·02) $\qquad \sigma(\infty) - \sigma(-\infty) = \sqrt{2\pi}\,\phi(0)$

when and only when $f(x)$ belongs to S'.

For by (21·26) and the proof of X_{40}, there is a sequence of values of ϵ such that if $\epsilon \to 0$ through that sequence, for almost all u

(23·03)
$$\sigma(u) - \sigma(-u) = \lim_{\epsilon \to 0} \frac{1}{2\epsilon\sqrt{2\pi}} \int_{-u}^{u} |s(u+\epsilon) - s(u-\epsilon)|^2 du,$$

and hence
$$\sigma(u) - \sigma(-u) \leqslant \lim_{\epsilon \to 0} \frac{1}{2\epsilon\sqrt{2\pi}} \int_{-\infty}^{\infty} |s(u+\epsilon) - s(u-\epsilon)|^2 du$$
$$= \sqrt{2\pi}\,\phi(0).$$

Thus, by the monotoneness of $\sigma(u)$,
$$\sigma(u) - \sigma(-u) \leqslant \sqrt{2\pi}\,\phi(0)$$
for *all* u, which establishes (23·01).

If (23·02) is to hold, we must have

(23·04) $\qquad \displaystyle\lim_{u \to \infty} \overline{\lim_{\epsilon \to 0}} \frac{1}{4\pi\epsilon} \left[\int_{u}^{\infty} + \int_{-\infty}^{-u} \right] |s(u+\epsilon) - s(u-\epsilon)|^2 du = 0$

at least if $\epsilon \to 0$ through our selected sequence. In the proof of the appropriate part of theorem 28, nothing is changed if we restrict ϵ to such a sequence. Thus $f(x)$ belongs to S'. On the other hand, if $f(x)$ belongs to S', by the other part of theorem 28 it will follow that (23·04) is true as $\epsilon \to 0$ by *any* route, and by (23·03) and (21·17), we arrive at (23·02).

Theorem 32. *If $f(x)$ belongs to S and we define $\sigma(u)$ as in lemma 30_a, and $\{u_n\}$ are the points where jumps of $\sigma(u)$ occur,*

$$\lim_{T \to \infty} \frac{1}{2T} \int_{-T}^{T} |\phi(x)|^2 dx = \frac{1}{2\pi} \Sigma \left[\sigma(u_n + 0) - \sigma(u_n - 0)\right]^2.$$

182   GENERALIZED HARMONIC ANALYSIS

In particular, if $\sigma(u)$ is continuous,

$$\lim_{T\to\infty}\frac{1}{2T}\int_{-T}^{T}|\phi(x)|^2\,dx=0.$$

This follows from lemma 30_a and theorem 24 and the fact that σ bears the same relation to ϕ as s does to f.

THEOREM 33. *Under the hypothesis of theorem 32, if we put*

(23·05) $\quad \psi(x)=\phi(x)-\Sigma\,\dfrac{1}{(2\pi)^{\frac{1}{2}}}\,e^{iu_n x}\left[\sigma(u_n+0)-\sigma(u_n-0)\right],$

then

(23·06) $\quad\quad\quad \lim\limits_{T\to\infty}\dfrac{1}{2T}\int_{-T}^{T}|\psi(x)|^2\,dx=0.$

The series in (23·06) is absolutely convergent by (23·01) and the monotoneness of $\sigma(u)$. We have

$$\frac{1}{\sqrt{2\pi}}\left\{\underset{A\to\infty}{\text{l.i.m.}}\left[\int_1^A+\int_{-A}^{-1}\right]\frac{\psi(\xi)\,e^{-iu\xi}}{-i\xi}\,d\xi+\int_{-1}^{1}\psi(\xi)\frac{e^{-iux}-1}{-i\xi}\,d\xi\right\}$$

$=\sigma(u)+\text{const.}-\tfrac{1}{2}\Sigma\left[\sigma(u_n+0)-\sigma(u_n-0)\right]\left[\text{sgn}\,u_n-\text{sgn}(u_n-u)\right],$

and this is a continuous monotone function. Then (22·06) follows by theorem 24.

THEOREM 34. *If $f(x)$ belongs to S, then, almost everywhere,*

(23·07) $\quad\quad \phi(x)=\dfrac{1}{\sqrt{2\pi}}\int_{-\infty}^{\infty}e^{iux}\,d\sigma(u).$

To begin with, the integral in (23·07) exists and is continuous and bounded for every real u, by (23·01) and the monotoneness of $\sigma(u)$. As for the continuity,

$$\left|\frac{1}{\sqrt{2\pi}}\int_{-\infty}^{\infty}(e^{iu(x+\delta)}-e^{iux})\,d\sigma(u)\right|\leqslant\sqrt{\frac{2}{\pi}}\int_{-\infty}^{\infty}\left|\sin\frac{\delta u}{2}\right|\,d\sigma(u)$$

$$\leqslant\sqrt{\frac{2}{\pi}}\left\{\left[\int_{\delta^{-\frac{1}{2}}}^{\infty}+\int_{-\infty}^{-\delta^{-\frac{1}{2}}}\right]d\sigma(u)+\frac{\delta^{\frac{1}{2}}}{2}\int_{-\infty}^{\infty}d\sigma(u)\right\},$$

and this can be made as small as we wish by taking δ small enough.

Let us put $\quad \phi_1(x)=\dfrac{1}{\sqrt{2\pi}}\int_{-\infty}^{\infty}e^{iux}\,d\sigma(u).$

Then
$$\phi_1(x)\frac{\sin \lambda x}{x} = \frac{1}{\sqrt{2\pi}}\int_{-\infty}^{\infty} e^{iux}\,d\sigma(u)\,\tfrac{1}{2}\int_{-\lambda}^{\lambda} e^{ivx}\,dv$$
$$= \frac{1}{\sqrt{2\pi}}\int_{-\infty}^{\infty} d\sigma(u)\,\tfrac{1}{2}\int_{-\lambda}^{\lambda} e^{i(u+v)x}\,dv$$
$$= \frac{1}{\sqrt{2\pi}}\int_{-\infty}^{\infty} e^{iwx}\,dw\,\tfrac{1}{2}\int_{w-\lambda}^{w+\lambda} d\sigma(u)$$
$$= \frac{1}{\sqrt{2\pi}}\int_{-\infty}^{\infty} e^{iwx}\,dw\,\tfrac{1}{2}[\sigma(w+\lambda)-\sigma(w-\lambda)].$$

Here all the inversions of the order of integration are justified by absolute convergence. Again, by (21·21),

$$\tfrac{1}{2}[\sigma(w+\lambda) - \sigma(w-\lambda)] = \frac{1}{\sqrt{2\pi}}\underset{A\to\infty}{\text{l.i.m.}}\int_{-A}^{A} e^{-iwx}\phi(x)\frac{\sin \lambda x}{x}\,dx.$$

As ϕ is bounded, $\phi(x)\sin \lambda x/x$ belongs to L_2, and, by the Plancherel theorem,

$$\phi(x)\frac{\sin \lambda x}{x} = \frac{1}{\sqrt{2\pi}}\underset{A\to\infty}{\text{l.i.m.}}\int_{-A}^{A} e^{iwx}\,dw\,\tfrac{1}{2}[\sigma(w+\lambda)-\sigma(w-\lambda)].$$

By X_{41}, it follows that, almost everywhere,

$$\phi(x)\frac{\sin \lambda x}{x} = \phi_1(x)\frac{\sin \lambda x}{x},$$

or that $\phi(x) = \phi_1(x)$ almost everywhere. This completes the proof of theorem 34. We have incidentally proved:

THEOREM 35. *If $f(x)$ belongs to S, then $\phi(x)$ differs from a continuous function at most over a null set of arguments.*

THEOREM 36. *If $f(x)$ belongs to S, then*
$$S(u) = \frac{1}{\sqrt{2\pi}}\int_{-\infty}^{\infty} \phi(x)\frac{e^{-iux}-1}{-ix}\,dx$$
exists for every u. We shall have
$$S(u) - \sigma(u) = \text{const.}$$
except over a null set.

184 GENERALIZED HARMONIC ANALYSIS

We shall take $\sigma(u)$ to be monotone and shall define it to be $\dfrac{\sigma(u+0)+\sigma(u-0)}{2}$ at points of discontinuity. We have

$$\frac{1}{\sqrt{2\pi}}\int_{-A}^{A}\phi(x)\frac{e^{-iux}-1}{-ix}dx$$

$$=\frac{1}{\sqrt{2\pi}}\int_{-A}^{A}\frac{e^{-iux}-1}{-ix}dx\,\frac{1}{\sqrt{2\pi}}\int_{-\infty}^{\infty}e^{iwx}d\sigma(w)$$

$$=\frac{1}{2\pi}\int_{-\infty}^{\infty}d\sigma(w)\int_{-A}^{A}\frac{e^{i(w-u)x}-e^{iwx}}{-ix}dx.$$

If we now integrate by parts, we shall find that the unintegrated terms vanish by the Riemann-Lebesgue theorem, and that

$$\frac{1}{\sqrt{2\pi}}\int_{-A}^{A}\phi(x)\frac{e^{-iux}-1}{-ix}dx$$

$$=\frac{-1}{2\pi}\int_{-\infty}^{\infty}\sigma(w)\,dw\int_{-A}^{A}\frac{d}{dw}\left(\frac{e^{i(w-u)x}-e^{iwx}}{-ix}\right)dx$$

$$=\frac{1}{2\pi}\int_{-\infty}^{\infty}\sigma(w)\,dw\int_{-A}^{A}(e^{i(w-u)x}-e^{iwx})dx$$

$$=\frac{1}{\pi}\int_{-\infty}^{\infty}\sigma(w)\left[\frac{\sin(w-u)A}{w-u}-\frac{\sin wA}{w}\right]dw$$

$$=\sigma(u)-\sigma(0)$$

$$+\frac{2}{\pi}\int_{0}^{\infty}\left[\frac{\sigma(u+w)-\sigma(w)+\sigma(u-w)-\sigma(-w)-2\sigma(u)+2\sigma(0)}{2}\right]$$
$$\times\frac{\sin wA}{w}dw.$$

Thus

$$(23\cdot 08)\quad \left|\frac{1}{\sqrt{2\pi}}\int_{-A}^{A}\phi(x)\frac{e^{-iux}-1}{-ix}dx-\sigma(u)+\sigma(0)\right|$$

$$=\left|\frac{2}{\pi}\int_{0}^{\infty}F(w)\frac{\sin wA}{w}dw\right|,$$

where $F(w)$ is a function vanishing continuously for $w=0$ and of limited total variation V over $(0,\infty)$. Then

$$(23\cdot 09)\quad \left|\frac{2}{\pi}\int_{0}^{\infty}F(w)\frac{\sin wA}{w}dw\right|=\left|\frac{2}{\pi}\int_{0}^{\infty}F\left(\frac{w}{A}\right)\frac{\sin w}{w}dw\right|$$

$$\leqslant 2n\limsup_{0<u\leqslant\frac{n\pi}{A}}|F(u)|+\left|\frac{2}{\pi}\int_{n\pi}^{\infty}F\left(\frac{w}{A}\right)\frac{\sin w}{w}dw\right|.$$

Again,

(23·10) $\left| \dfrac{2}{\pi} \displaystyle\int_{n\pi}^{\infty} F\left(\dfrac{w}{A}\right) \dfrac{\sin w}{w} dw \right|$

$\leqslant \left| \dfrac{2}{\pi} \displaystyle\int_{n\pi}^{\infty} \left[\displaystyle\int_{n\pi}^{w} \dfrac{\sin v}{v} dv \right] dF\left(\dfrac{w}{A}\right) \right|$

$+ \dfrac{2}{\pi} \limsup \left| F\left(\dfrac{w}{A}\right) \right| \left| \displaystyle\int_{n\pi}^{\infty} \dfrac{\sin v}{v} dv \right|$

$\leqslant \dfrac{4V}{\pi} \displaystyle\int_{n\pi}^{(n+1)\pi} \dfrac{|\sin v|}{v} dv$

$\leqslant 4V/n\pi.$

Here we have made use of the fact that F is bounded, and that $\dfrac{\sin w}{w}$ vanishes at $n\pi$ and at ∞. Now let n be so large that

$$4V/n\pi < \epsilon/2,$$

and then let A be so large that

$$2n \limsup_{0 < u \leqslant \frac{n\pi}{A}} |F(u)| < \epsilon/2.$$

It will follow from (23·08–10) that

$$\left| \dfrac{1}{\sqrt{2\pi}} \int_{-A}^{A} \phi(x) \dfrac{e^{-iux} - 1}{-ix} dx - \sigma(u) + \sigma(0) \right| < \epsilon.$$

Hence $\quad \dfrac{1}{\sqrt{2\pi}} \displaystyle\int_{-\infty}^{\infty} \phi(x) \dfrac{e^{-iux} - 1}{-ix} dx = \sigma(u) - \sigma(0),$

which is equivalent to the conclusion of theorem 35.

§ 24. The Elementary Properties of Almost Periodic Functions.

Let $f(x)$ be a continuous (real or complex) function of the real argument x, which ranges over $(-\infty, \infty)$. Following H. Bohr[*], we shall call $\tau(\epsilon)$ a *translation number* of $f(x)$ pertaining to ϵ in case for all x,

(24·01) $\qquad |f(x + \tau(\epsilon)) - f(x)| \leqslant \epsilon.$

We shall say that $f(x)$ is *uniformly almost periodic* in case there exists for every $\epsilon < 0$ a number $L(\epsilon)$ with the property that every interval $(A, A + L(\epsilon))$ contains at least one translation number $\tau(\epsilon)$ of $f(x)$ pertaining to ϵ. The literature[†] contains

[*] Bohr 1. [†] Besicovitch 1.

more general classes of almost periodic functions than the uniformly almost periodic functions, but we shall not consider them in the present book, and shall call the uniformly almost periodic functions "almost periodic" *tout net*. The principal theorems of Bohr on uniformly almost periodic functions are:

Theorem 37. *The class of almost periodic functions is identical with the class of functions $f(x)$ with the property that if $\epsilon < 0$, there exists a finite set of complex numbers A_k and real numbers Λ_k ($1 \leq k \leq n$), such that, for all x,*

(24·02) $$\left| f(x) - \sum_{1}^{n} A_k e^{i\Lambda_k x} \right| \leq \epsilon.$$

Theorem 38. *If $f(x)$ is almost periodic,*

(24·03) $$\lim_{T \to \infty} \frac{1}{2T} \int_{-T}^{T} f(x) e^{-i\Lambda x} dx$$

exists for every real Λ. There are at most a denumerable set of values of Λ for which the expression in (24·03) is unequal to zero. Let us call these $\Lambda_1, \Lambda_2, \ldots$. Let us put

$$A_n = \lim_{T \to \infty} \frac{1}{2T} \int_{-T}^{T} f(x) e^{-i\Lambda_n x} dx.$$

Then $$\lim_{n \to \infty} \lim_{T \to \infty} \frac{1}{2T} \int_{-T}^{T} \left| f(x) - \sum_{1}^{n} A_k e^{i\Lambda_k x} \right|^2 dx = 0.$$

These are respectively known as the Weierstrass and the Parseval theorems for almost periodic functions—the former from its analogy to the ordinary Weierstrass theorem on polynomial approximation, and the latter from its analogy to X_{51}. We shall devote the remainder of this book to their proof by means of a series of lemmas.

It will be seen that an ordinary periodic function will have its periods as translation numbers pertaining to every ϵ, and that every interval $L(\epsilon)$ of length exceeding the smallest positive period will contain at least one period. We thus obtain:

Lemma 37_1. *Every continuous periodic function is almost periodic.*

This serves as a justification of the name, "almost periodic."

If $f(x)$ is almost periodic, it is continuous, and will hence be less in modulus than some number N over the finite interval $(0, L(\epsilon))$. However, if x is any real number, there is a translation number $\tau(\epsilon)$ of $f(x)$ pertaining to ϵ and lying in $(x - L(\epsilon), x)$. Thus $x - \tau(\epsilon)$ lies in $(0, L(\epsilon))$, and, by (24·01),

$$|f(x)| \leqslant |f(x - \tau(\epsilon))| + \epsilon \leqslant N + \epsilon.$$

Hence we have:

Lemma 37₂. *Every almost periodic function is bounded.*

In exactly the same way, we may prove:

Lemma 37₃. *Every almost periodic function is uniformly continuous.*

For let $\epsilon > 0$. Let δ be so small that over the interval

$$(-\delta, L(\epsilon/3) + \delta), \quad |f(x) - f(y)| \leqslant \epsilon/3$$

whenever $|x - y| \leqslant \delta$. Let ξ and η be any two real numbers not differing by more than δ. Then there is some translation number $\tau(\epsilon/3)$ such that $\xi - \tau(\epsilon/3)$ and $\eta - \tau(\epsilon/3)$ both lie in $(-\delta, L(\epsilon/3) + \delta)$. Thus

$$|f(\xi) - f(\eta)| \leqslant |f(\xi) - f(\xi - \tau(\epsilon/3))| + |f(\eta) - f(\eta - \tau(\epsilon/3))|$$
$$+ |f(\xi - \tau(\epsilon/3)) - f(\eta - \tau(\epsilon/3))|$$
$$\leqslant \epsilon/3 + \epsilon/3 + \epsilon/3 = \epsilon.$$

A most important lemma is:

Lemma 37₄. *Let $f(x)$ and $g(x)$ be almost periodic. Then if $\epsilon > 0$, there is a number $M(\epsilon)$ such that every interval $(A, A + M(\epsilon))$ contains at least one number $\psi(\epsilon)$ that is a translation number pertaining to ϵ of both $f(x)$ and $g(x)$.*

To prove this, let $L_1(\epsilon)$ be a number such that every interval $(A, A + L_1(\epsilon))$ contains at least one translation number of $f(x)$ pertaining to ϵ, and let $L_2(\epsilon)$ bear the same relation to $g(x)$. Let δ_1 be so small that for $|x - y| \leqslant \delta_1$,

(24·04) $$|f(x) - f(y)| \leqslant \epsilon,$$

and let δ_2 bear a similar relation to $g(x)$. Let us divide the interval $(0, L_1(\epsilon))$ into m consecutive intervals I_k of size not exceeding δ_1, and similarly let us divide $(0, L_2(\epsilon))$ into n consecutive intervals J_k of size not exceeding δ_2. If x is any real number, there will be a translation number $\tau_1(\epsilon)$ of $f(x)$ pertaining to ϵ, lying between $x - L_1(\epsilon)$ and x. If we divide this by $L_1(\epsilon)$, let the least positive (or zero) remainder be ξ. Let ξ lie in I_μ. Similarly, let $\tau_2(\epsilon)$ be a translation number of $g(x)$ pertaining to ϵ, lying between $x - L_2(\epsilon)$ and x. If we divide this by $L_2(\epsilon)$, let the least positive (or zero) remainder be η. Let η lie in J_ν. We shall call (μ, ν) the *index* of x. There are not more than mn possible indices.

In some finite interval, say $(-X, X)$, x will run through all the indices it ever attains. Thus if x is any number whatever, there is a number y in $(-X, X)$ with the same index (μ, ν). This implies that there lie in I_μ two numbers, say a and b, such that a differs from x by a translation number of $f(x)$ pertaining to ϵ and that b differs from y by a translation number pertaining to ϵ. That is, if λ is any quantity,

(24·05) $\quad |f(a+\lambda) - f(x+\lambda)| \leqslant \epsilon; \quad |f(b+\lambda) - f(y+\lambda)| \leqslant \epsilon.$

However, $\quad |(a+\lambda) - (b+\lambda)| \leqslant \delta_1,$

since a and b lie in the same I_μ. Thus, by (24·04),

$$|f(a+\lambda) - f(b+\lambda)| \leqslant \epsilon.$$

Also, by (24·05), $\quad |f(x+\lambda) - f(y+\lambda)| \leqslant 3\epsilon.$

Similarly, $\quad |g(x+\lambda) - g(y+\lambda)| \leqslant 3\epsilon.$

Thus $x - y$ is a translation number pertaining to 3ϵ of both $f(x)$ and $g(x)$. Since y lies somewhere in $(-X, X)$ for an arbitrary x, it follows that every interval of length $2X$ contains at least one common translation number of $f(x)$ and $g(x)$ pertaining to 3ϵ. Since 3ϵ is an arbitrary positive number, this establishes lemma 37_4.

As an immediate corollary, we have:

Lemma 37_5. *If $f(x)$ and $g(x)$ are almost periodic, so are $f(x) \pm g(x)$ and $f(x) g(x)$.*

For any translation number pertaining to ϵ for both $f(x)$ and $g(x)$ is a translation number of $f(x) \pm g(x)$ pertaining to 2ϵ, and a translation number of $f(x) g(x)$ pertaining to

$$\epsilon \{\limsup |f(x)| + \limsup |g(x)|\}.$$

In particular, if $f(x)$ is almost periodic, so is $f(x) e^{i\lambda x}$.

GENERALIZED HARMONIC ANALYSIS   189

A very important lemma of the same type is:

Lemma 37₆. *If $f(x)$ is almost periodic, and if $F(u)$ is a continuous function of u over a closed interval containing the range of values of $f(x)$, then $F(f(x))$ is almost periodic.*

For if $|F(u_2) - F(u_1)|$ is less than ϵ whenever $|u_2 - u_1| \leq \delta$, every translation number of $f(x)$ pertaining to δ is a translation number of $F(f(x))$ pertaining to ϵ.

A trivial instance is that $\bar{f}(x)$ is almost periodic when $f(x)$ is. Since obviously $f(x+\lambda)$ is almost periodic, it follows by lemma 37₅ that $f(x+\lambda)\bar{f}(x)$ is almost periodic in x. Again, $|f(x)|$ will be almost periodic when $f(x)$ is.

Lemma 37₇. *If $f(x)$ is almost periodic, there exists a finite number A such that*

$$(24{\cdot}055) \qquad A = \lim_{T \to \infty} \frac{1}{T} \int_y^{y+T} f(x)\,dx$$

uniformly in y.

Let $\epsilon > 0$. Let U be some translation number of $f(x)$ pertaining to ϵ, and exceeding $L(\epsilon)$—say the least such number. Then there always is a translation number V pertaining to ϵ between nU and $(n+1)U$. Hence

$$\left| \frac{1}{U}\int_{nU}^{(n+1)U} f(x)\,dx - \frac{1}{U}\int_0^U f(x)\,dx \right| = \frac{1}{U}\left| \int_{nU}^V f(x)\,dx - \int_{nU-V}^0 f(x)\,dx \right.$$

$$+ \int_{nU-V}^0 f(x)\,dx - \int_{(n+1)U-V}^U f(x)\,dx$$

$$\left. + \int_V^{(n+1)U} f(x)\,dx - \int_0^{(n+1)U-V} f(x)\,dx \right|$$

$$\leq \frac{\epsilon}{U}[2V - 2nU + (n+1)U - V]$$

$$= \frac{\epsilon}{U}(V - nU + U)$$

$$\leq 2\epsilon.$$

Hence, if μ and ν are integers,

$$(24{\cdot}06) \qquad \left| \frac{1}{(\mu-\nu)U} \int_{\nu U}^{\mu U} f(x)\,dx - \frac{1}{U}\int_0^U f(x)\,dx \right| \leq 2\epsilon.$$

190   GENERALIZED HARMONIC ANALYSIS

Now let $\nu U \leqslant y \leqslant (\nu+1)U$ and $\mu U \leqslant y+T \leqslant (\mu+1)\,U$. Then

$$(24\cdot 07) \quad \left| \frac{1}{T} \int_y^{y+T} f(x)\,dx - \frac{1}{(\mu-\nu)U} \int_{\nu U}^{\mu U} f(x)\,dx \right|$$

$$\leqslant \left| \frac{1}{T} \int_y^{\nu U} f(x)\,dx \right| + \left| \frac{1}{T} \int_{\mu U}^{y+T} f(x)\,dx \right| + \left| \left(\frac{1}{T} - \frac{1}{(\mu-\nu)U}\right) \int_{\nu U}^{\mu U} f(x)\,dx \right|$$

$$\leqslant \frac{2U}{T} \limsup |f(x)| + \left| (\mu-\nu)U \left[\frac{1}{(\mu-\nu)U} - \frac{1}{T} \right] \right|$$

$$\times \left[2\epsilon + \left| \frac{1}{U} \int_0^U f(x)\,dx \right| \right]$$

$$\leqslant \frac{3U}{T} \limsup |f(x)| + \frac{2\epsilon U}{T}.$$

Here we have applied (24·06) in order to obtain an estimate of

$$\frac{1}{(\mu-\nu)U} \int_{\nu U}^{\mu U} f(x)\,dx.$$

It follows from (24·07) that we may choose T so large independently of y that, for some μ and ν,

$$\left| \frac{1}{T} \int_y^{y+T} f(x)\,dx - \frac{1}{(\mu-\nu)U} \int_{\nu U}^{\mu U} f(x)\,dx \right| \leqslant \epsilon.$$

Thus, by (24·06), for all y,

$$(24\cdot 08) \quad \left| \frac{1}{T} \int_y^{y+T} f(x)\,dx - \frac{1}{U} \int_0^U f(x)\,dx \right| \leqslant 3\epsilon.$$

Hence, if $f(x)$ is real,

$$\overline{\lim_{T\to\infty}} \frac{1}{T} \int_y^{y+T} f(x)\,dx - \underline{\lim_{T\to\infty}} \frac{1}{T} \int_y^{y+T} f(x)\,dx \leqslant 6\epsilon,$$

and, since ϵ is arbitrary,

$$\overline{\lim_{T\to\infty}} \frac{1}{T} \int_y^{y+T} f(x)\,dx = \underline{\lim_{T\to\infty}} \frac{1}{T} \int_y^{y+T} f(x)\,dx = \lim_{T\to\infty} \frac{1}{T} \int_y^{y+T} f(x)\,dx.$$

The existence of this limit, which we shall write A, follows even in the complex case from a separate consideration of real and imaginary parts. Let us note that A is independent of y, and that y may go through any real set of values as T becomes infinite without vitiating (24·055). This completes the proof of lemma 37$_7$.

GENERALIZED HARMONIC ANALYSIS

As a corollary, we see that if we put*

$$\lim_{T \to \infty} \frac{1}{T} \int_y^{y+T} f(x)\, dx = M\{f(x)\} = M_x\{f\},$$

then if $f(x)$ is almost periodic, there exist uniformly in y:
$M\{f(x)e^{i\lambda x}\}$ for all real λ; $M_\xi\{f(x+\xi)\bar{f}(\xi)\}$; $M\{|f(x)|^2\}$.

Lemma 37₈. *If $f(x)$ is almost periodic, so is*

$$\phi(x) = M_\xi\{f(x+\xi)\bar{f}(\xi)\}.$$

Clearly

$$|\phi(x+\tau) - \phi(x)| = |M_\xi\{[f(x+\xi+\tau) - f(x+\xi)]\bar{f}(x)\}|$$
$$\leqslant \limsup |f(\xi)| \limsup |f(\xi+\tau) - f(\xi)|.$$

Thus any translation number of $f(x)$ pertaining to ϵ is a translation number of $\phi(x)$ pertaining to $\epsilon \limsup |f(\xi)|$. Lemma 37₈ follows at once.

Lemma 37₉. *If $f(x)$ is a non-negative almost periodic function and*

(24·09) $$M\{f(x)\} = 0,$$

then $f(x)$ vanishes identically.

Otherwise there will be a point x such that $f(x) > 0$. Let $f(x) = 2\eta$. Because of the continuity of $f(x)$, there will then be an interval (a, b) about x in which $f(x) \geqslant \eta$. Let $L(\eta/2)$ be such that every interval of length $L(\eta/2)$ contains at least one translation number of f pertaining to $\eta/2$. It will then follow that every interval of length $L(\eta/2)$ will contain an interval of length $(b - a)$ which is transformed into (a, b) by such a translation number, so that in this interval, $f(x) \geqslant \eta/2$. That is, if

$$nL(\xi/2) \leqslant T \leqslant (n+1) L(\eta/2),$$

we shall have

$$\frac{1}{T}\int_y^{y+T} f(x)\, dx \geqslant \frac{n-1}{n+1} \frac{(b-a)\eta}{2L(\eta/2)},$$

so that

$$\lim_{T \to \infty} \frac{1}{T} \int_y^{y+T} f(x)\, dx \geqslant \frac{(b-a)\eta}{2L(\eta/2)} > 0.$$

---

* We use the M_x notation when there is any possible ambiguity as to the variable with respect to which we are taking the mean.

192 GENERALIZED HARMONIC ANALYSIS

This contradicts the hypothesis (24·09), so that no such point x exists and $f(x)$ vanishes identically.

Lemma 37₁₀. *If $f(x)$ is an almost periodic function,*
$$(24\cdot 10) \qquad g(x) = \limsup_{-\infty < y < \infty} |f(x+y) - f(y)|$$
is almost periodic*.

For
$$|g(x+\tau) - g(x)| = \left| \limsup_{-\infty < y < \infty} |f(x+\tau+y) - f(y+\tau)| \right.$$
$$\left. - \limsup_{-\infty < y < \infty} |f(x+y) - f(y)| \right|$$
$$\leqslant \limsup_{-\infty < y < \infty} \left| |f(x+\tau+y) - f(y+\tau)| - |f(x+y) - f(y)| \right|$$
$$\leqslant \limsup_{-\infty < y < \infty} |f(x+\tau+y) - f(x+y) - f(y+\tau) + f(y)|$$
$$\leqslant \limsup_{-\infty < y < \infty} \{|f(x+\tau+y) - f(x+y)| + |f(y+\tau) - f(y)|\}$$
$$\leqslant 2 \limsup_{-\infty < y < \infty} |f(y+\tau) - f(y)|.$$

Thus every translation number of $f(x)$ pertaining to ϵ will be a translation number of $g(x)$ pertaining to 2ϵ. This establishes lemma 37₁₀.

Lemma 37₁₁. *Let $f(x)$ be almost periodic. Let $g(x)$ be defined as in (24·10). Let $G_\epsilon(x)$ be defined by*
$$G_\epsilon(x) = 1 - \frac{x}{\epsilon} \quad (0 < x \leqslant \epsilon); \qquad G_\epsilon(x) = 0 \quad (\epsilon \leqslant x).$$
Then $G_\epsilon(g(x))$ will be non-negative, almost periodic, and not identically zero. If
$$G_\epsilon(g(\tau)) \neq 0,$$
τ will be a translation number of $f(x)$ pertaining to ϵ. We shall put
$$(24\cdot 11) \qquad \psi_\epsilon(x) = G_\epsilon(g(x))/M\{G_\epsilon(g(y))\}.$$
Then $\psi_\epsilon(x)$ will exist.

The almost-periodicity of $G_\epsilon(g(x))$ will result from lemmas 37₆ and 37₁₀. The non-negativeness is a matter of definition. $G_\epsilon(g(x))$

* Cf. Bochner 1.

GENERALIZED HARMONIC ANALYSIS

cannot vanish for $x=0$. If $G_\epsilon(g(x)) \neq 0$, we have $g(x) < \epsilon$, and hence, by (24·10),
$$|f(\tau+y)-f(y)| < \epsilon$$
for all y, so that τ is a translation number of $f(x)$ pertaining to ϵ. By lemma 37_9, $M\{G_\epsilon(g(y))\} \neq 0$, and $\psi_\epsilon(x)$ is properly defined by (24·11).

Lemma 37_{12}. *If $f(x)$ is almost periodic and $\psi_\epsilon(x)$ is defined by* (24·11),

(24·12) $$f_\epsilon(x) = M_y\{\psi_\epsilon(x-y)f(y)\}$$
exists and is almost periodic, and

(24·13) $$|f(x) - f_\epsilon(x)| \leqslant \epsilon$$
for every x. Furthermore,

(24·14) $$_\epsilon f(x) = M_z\{\psi_\epsilon(x-z)f_\epsilon(z)\}$$
exists and is almost periodic, and

(24·15) $$|_\epsilon f(x) - f_\epsilon(x)| \leqslant \epsilon$$
for all values of x, so that

(24·16) $$|_\epsilon f(x) - f(x)| \leqslant 2\epsilon.$$

We have

(24·17) $$_\epsilon f(x) = M_y\{f(y) M_z\{\psi_\epsilon(z+x-y) \psi_\epsilon(z)\}\}.$$

To begin with, by the definition of $\psi_\epsilon(x)$ and the fact that $\psi(x)$ differs from 0 only for arguments that are translation numbers of $f(x)$ pertaining to ϵ, $f_\epsilon(x)$ is a weighted average with positive weighting of quantities like $f(x+\tau(\epsilon))$, which differ from $f(x)$ by not more than ϵ. We thus get (24·13). Moreover, by lemmas 37_7 and 37_5,
$$f_\epsilon(x) = M_y\{f(x+y) \psi_\epsilon(x)\},$$
and, since by (24·11),
$$M\{\psi_\epsilon(x)\} = 1,$$
it is easy to see that
$$\limsup_{-\infty < x < \infty} |f_\epsilon(x+\tau) - f_\epsilon(x)| \leqslant \limsup_{-\infty < x < \infty} |f(x+\tau) - f(x)|,$$

194 GENERALIZED HARMONIC ANALYSIS

and that $f_\epsilon(x)$ has the same translation numbers pertaining to the same quantities as does $f(x)$. Thus $f_\epsilon(x)$ is almost periodic, (24·14) exists and is almost periodic by the same arguments which we have already used in the case of (24·12), and (24·15) is true for the same reasons as (24·13). If we put (24·13) and (24·15) together, we get (24·16).

As to (24·17), we have, by (24·14) and (24·12),

$$_\epsilon f(x) = M_z \{\psi_\epsilon(x-z) M_y \{\psi_\epsilon(z-y) f(y)\}\}$$

$$= \lim_{T_1 \to \infty} \frac{1}{T_1} \int_a^{a+T_1} \psi_\epsilon(x-z)\, dz \lim_{T_2 \to \infty} \frac{1}{T_2}$$

$$\times \int_\beta^{\beta+T_2} \psi_\epsilon(z-y) f(y)\, dy.$$

Since ψ_ϵ and f are almost periodic and hence bounded, the expression under the second limit sign tends to its limit boundedly, and, by bounded convergence,

$$_\epsilon f(x) = \lim_{T_1 \to \infty} \lim_{T_2 \to \infty} \frac{1}{T_1 T_2} \int_a^{a+T_1} \psi_\epsilon(x-z)\, dz$$

$$\times \int_\beta^{\beta+T_2} \psi_\epsilon(z-y) f(y)\, dx$$

$$= \lim_{T_1 \to \infty} \lim_{T_2 \to \infty} \frac{1}{T_2} \int_\beta^{\beta+T_2} f(y)\, dy \frac{1}{T_1}$$

$$\times \int_a^{a+T_1} \psi_\epsilon(x-z) \psi_\epsilon(z-y)\, dz,$$

or, since the function ψ_ϵ is even,

(24·18)

$$_\epsilon f(x) = \lim_{T_1 \to \infty} \lim_{T_2 \to \infty} \frac{1}{T_2} \int_\beta^{\beta+T_2} f(y)\, dy \frac{1}{T_1}$$

$$\times \int_a^{a+T_1} \psi_\epsilon(z-y) \psi_\epsilon(z-x)\, dz$$

$$= \lim_{T_1 \to \infty} \lim_{T_2 \to \infty} \frac{1}{T_2} \int_\beta^{\beta+T_2} f(y)\, dy \frac{1}{T_1}$$

$$\times \int_{a-x}^{a-x+T_1} \psi_\epsilon(z+x-y) \psi_\epsilon(z)\, dz$$

$$= \lim_{T_1 \to \infty} M_y \left\{ f(y) \frac{1}{T_1} \int_{a-x}^{a-x+T_1} \psi_\epsilon(z+x-y) \psi_\epsilon(z)\, dz \right\}.$$

If we can show that
(24·19)
$$\lim_{T_1 \to \infty} \frac{1}{T_1} \int_{\gamma}^{\gamma+T_1} \psi_\epsilon(z+u)\, \psi_\epsilon(z)\, dz = M_z \{\psi_\epsilon(z+u)\, \psi_\epsilon(z)\}$$
uniformly in u and γ, (24·17) will be an immediate consequence of (24·18). However, in the argument of lemma 37_7, when we showed (24·055) to exist uniformly in y, our proof gave an estimate of this uniform approach to a limit which involved solely:

(i) The set of translation numbers of $f(x)$ for all different values of ϵ;

(ii) The least upper bound of $|f(x)|$.

Now, the functions $\psi_\epsilon(z+u)\, \psi_\epsilon(z)$ have as a common upper bound for their modulus the square of the upper bound of the modulus of $\psi_\epsilon(z)$. Furthermore, any translation number of $\psi_\epsilon(z)$ pertaining to ϵ is a translation number of $\psi_\epsilon(z+u)\, \psi_\epsilon(z)$ for *every* u pertaining to
$$2\epsilon \limsup_{-\infty < z < \infty} |\psi_\epsilon(z)|.$$
We may thus establish the following lemma just as we have established lemma 37_7.

Lemma 37_{13}. *If $f(x)$ is almost periodic, (24·19) is true uniformly of γ and u.*

This completes the proof of lemma 37_{12}.

Lemma 37_{14}. *If $\{f_n(x)\}$ is a sequence of almost periodic functions, tending uniformly to a limit $f(x)$, then $f(x)$ is almost periodic. In particular, if $\Sigma |B_n| < \infty$, and $\{\Lambda_n\}$ is a set of real numbers,*

(24·20) $\qquad\qquad \Sigma B_n e^{i\Lambda_n x}$

is almost periodic.

For let $\qquad |f_n(x) - f(x)| \leqslant \epsilon/3$
for all x, and let $\tau(\epsilon)$ be a translation number of $f_n(x)$ pertaining to $\epsilon/3$. Then $\tau(\epsilon)$ will be a translation number of $f(x)$ per-

196   GENERALIZED HARMONIC ANALYSIS

taining to ϵ. Thus there will exist a number $L(\epsilon)$ such that each interval $(A, A + L(\epsilon))$ will contain at least one $\tau(\epsilon)$. As $f(x)$ is continuous, being the limit of a uniformly convergent sequence of continuous functions, lemma 37_{14} is established. By lemmas 37_1 and 37_5, it follows that $\sum_1^N B_n e^{i\Lambda_n x}$ is almost periodic, so that the almost periodicity of (24·20) follows at once.

§ 25. The Weierstrass and Parseval Theorems for Almost Periodic Functions.

Lemma 37_{15}. *If $f(x)$ is almost periodic, then there exists a (finite or denumerable) sequence of positive numbers B_n, with corresponding real numbers Λ_n, such that*

(25·01) $\qquad\qquad \sum B_n < \infty,$

and

(25·02) $\qquad \phi(x) = M_\xi \{f(x + \xi)\bar{f}(\xi)\} = \sum B_n e^{i\Lambda_n x}.$

Here we appeal to theorem 32. If $f(x)$ is almost periodic, so is $\phi(x)$ by lemma 37_8. By the boundedness and monotonic character of $\sigma(u)$,

$$\sum \frac{1}{(2\pi)^{\frac{1}{2}}} e^{iu_n x} [\sigma(u_n + 0) - \sigma(u_n - 0)]$$

is almost periodic in accordance with lemma 37_{14}. Thus by lemma 37_5, the function $\psi(x)$ of (23·05) is almost periodic. In view of (23·06) and lemma 37_9,

$$\psi(x) \equiv 0,$$

and $\qquad \phi(x) = \sum \dfrac{1}{(2\pi)^{\frac{1}{2}}} e^{iu_n x} [\sigma(u_n + 0) - \sigma(u_n - 0)].$

This is however an expression of the type of (25·02), and lemma 37_{15} is established.

We may now complete the proof of theorem 37. By lemma 37_{15}, (24·16), and (24·17), if $f(x)$ is almost periodic and $\epsilon > 0$, there exists an absolutely convergent series $\sum B_n$ and a sequence $\{\Lambda_n\}$ of real exponents, such that

(25·03) $\qquad |f(x) - M_y \{f(y) \sum B_n e^{i\Lambda_n(x-y)}\}| \leqslant 2\epsilon.$

GENERALIZED HARMONIC ANALYSIS

In view of the fact that $f(y)e^{-i\Lambda_n y}$ is uniformly bounded, either
$$\Sigma B_n e^{i\Lambda_n(x-y)}$$
contains a finite number of terms to begin with, or we may choose N so large that

(25·04)
$$\left| M_y\left\{f(y) \sum_1^N B_n e^{i\Lambda_n(x-y)}\right\} - M_y\left\{f(y) \sum_1^\infty B_n e^{i\Lambda_n(x-y)}\right\} \right| < \epsilon.$$

Combining (25·03) and (25·04), we get
$$\left| f(x) - M_y\left\{f(y) \sum_1^N B_n e^{i\Lambda_n(x-y)}\right\} \right| < 3\epsilon.$$

If we put $\quad A_n = B_n M_y\{f(y)e^{-i\Lambda_n y}\},$
this gives us

(25·05) $\qquad \left| f(x) - \sum_1^N A_n e^{i\Lambda_n x} \right| < 3\epsilon.$

Thus, for any ϵ, we may find complex coefficients A_k and real exponents Λ_k for which (24·02) is true, and we have established theorem 37.

As to theorem 38, let $\Lambda_1, \ldots, \Lambda_N$ be given, and let us seek to choose $A_1', \ldots, A_N'$ so as to minimize
$$M\left\{\left| f(x) - \sum_1^N A_n' e^{i\Lambda_n x}\right|^2\right\},$$
which exists as the mean of an almost periodic function. We shall have

(25·06)
$$M\left\{\left| f(x) - \sum_1^N A_n' e^{i\Lambda_n x}\right|^2\right\}$$
$$= M\{|f(x)|^2\} - \sum_1^N A_n' M\{\overline{f(x)} e^{i\Lambda_n x}\}$$
$$\quad - \sum_1^N \overline{A}_n' M\{f(x)e^{-i\Lambda_n x}\} + \sum_1^N |A_n'|^2$$
$$= M\{|f(x)|^2\} - \sum_1^N |M\{f(x)e^{-i\Lambda_n x}\}|^2$$
$$\quad + \sum_1^N |A_n' - M\{f(x)e^{-i\Lambda_n x}\}|^2.$$

This will assume its minimum value when, and only when,

(25·07) $\quad A_n' = M\{f(x) e^{-i\Lambda_n x}\}, \quad [n = 1, 2, \ldots, N]$.

We shall have

(25·08) $\quad M\{|f(x)|^2\} \geq \sum_1^N |M\{f(x) e^{-i\Lambda_n x}\}|^2$.

There can thus only be a finite number of values of Λ_n on $(-\infty, \infty)$ for which

$$|M\{f(x) e^{-i\Lambda_n x}\}| > a > 0,$$

and at most a denumerable set of values for which

$$M\{f(x) e^{-i\Lambda_n x}\} \neq 0.$$

Let the totality of these values be $\{\lambda_n\}$, and let

$$a_n = M\{f(x) e^{-i\lambda_n x}\}.$$

Then, by (25·08),

(25·09) $\quad \Sigma |a_n|^2 \leq M\{|f(x)|^2\}$.

On the other hand, by (25·05), if we define A_n' as in (25·07),

$$M\left\{\left|f(x) - \sum_1^N A_n' e^{i\Lambda_n x}\right|^2\right\} \leq M\left\{\left|f(x) - \sum_1^N A_n e^{i\Lambda_n x}\right|^2\right\}$$
$$\leq 9\epsilon^2.$$

By (25·06), $\quad M\{|f(x)|^2\} - \sum_1^N |A_n'|^2 \leq 9\epsilon^2$,

and since the A_n''s which differ from 0 represent a selection among the a_n's, and ϵ is arbitrary,

$$M\{|f(x)|^2\} \geq \Sigma |a_n|^2.$$

Thus, by (25·09),

$$M\{|f(x)|^2\} = \Sigma |a_n|^2.$$

In other words, by (25·06),

(25·10) $\quad \lim_{n \to \infty} M\left\{\left|f(x) - \sum_1^n a_n e^{i\lambda_n x}\right|^2\right\} = 0$,

and we have established theorem 37.

A historical remark concerning theorems 37 and 38 is perhaps in order. The first proof in each case is that of Bohr himself[*], and is technically elementary but by no means easy. The present

[*] Bohr 1.

author* gave the next proof, depending like the present one on the Fourier integral, but much more intricate. A further proof† was due to H. Weyl, who for the first time explicitly introduced the function $\phi(x)$ which had appeared in a modified form in the author's papers. Weyl's proof depends on the ideas and technique of the theory of integral equations, though not on the proved theorems in that field, and is to my mind the simplest and most direct way of proving theorems 37 and 38 if these theorems alone are required, but it does not orient them with respect to the more general theory of harmonic analysis. The introduction of $\psi_\epsilon(x)$ in the treatment of the present paragraph is an adaptation from Weyl.

The next proofs to appear were those of the present author‡, substantially as they are here presented. At about the same time, de la Vallée Poussin§ gave a proof depending on the same ideas as the original proof of Bohr, but also introducing the function $\phi(x)$. This proof has found favour at the hands of Bohr and Besicovitch, though the present author cannot altogether agree with them in preferring it to that of Weyl. Like Weyl's proof, it is not adapted to an extension to a more general theory of harmonic analysis

The dominating idea in proofs of the Bohr-de la Vallée Poussin type is that of the arrangement of the terms $a_n e^{i\lambda_n x}$ in (25·10) in an order depending on the arithmetical properties of the λ_n. The dominating idea in the Weyl proof is that of the arrangement of these terms in the descending order of magnitude of the coefficients $|a_n|$. The present treatment arranges these terms in the order of the exponents λ_n. This is the only order which is compatible with a unified treatment of almost periodic functions and functions with continuous spectra.

* Wiener 5.   † Weyl 2.
‡ Wiener 1, 6.   § De la Vallée Poussin 1.

BIBLIOGRAPHY

[The author has already given detailed bibliographies of generalized harmonic analysis in Wiener 1 and of Tauberian theorems in Wiener 2. Accordingly, the present bibliography is devoted exclusively to papers cited in the text.]

A. S. Besicovitch 1. *Almost Periodic Functions.* Cambridge Univ. Press, 1932, pp. xiii+180.

S. Bochner 1. "Beiträge zur Theorie der fastperiodischen Funktionen." *Math. Ann.* 96 (1926), 119-147.

H. Bohr 1. "Zur Theorie der fastperiodischen Funktionen." *Acta Math.* 45 (1924), 29-127.

H. E. Bray 1. "Elementary Properties of the Stieltjes Integral." *Annals of Mathematics*, 20 (1919), 176-186.

G. A. Campbell and R. M. Foster 1. "Fourier Integrals for Practical Applications." *Bell telephone syst. techn. publ., Math.-phys. Monogr.* B 584.

P. J. Daniell 1. "A General Form of Integral." *Annals of Mathematics*, 19 (1917), 279-294.

E. Fischer 1. *Comptes Rendus*, 144 (1907), 1022-1024.

G. H. Hardy and J. E. Littlewood 1. "On a Tauberian Theorem for Lambert's Series, and Some Fundamental Theorems in the Analytic Theory of Numbers." *Proc. Lond. Math. Soc.* (2), 19 (1921), 21-29.

―― ―― 2. "The Riemann Zeta Function and the Theory of the Distribution of the Primes." *Acta Math.* 41 (1918), 119-196.

S. Ikehara 1. "An extension of Landau's Theorem in the Analytic Theory of Numbers." *Journ. Math. Phys. M.I.T.* 10 (1931), 1-12.

J. Karamata 1. "Über die Hardy-Littlewoodschen Umkehrungen des Abelschen Stetigkeitssatzes." *Math. Zeitschr.* 32 (1930), 319-20.

K. Knopp 1. "Über Lambertsche Reihen." *Journ. f. d. reine u. angew. Mathem.* 142 (1913), 283-315.

E. Landau 1. "Über einen Satz des Herrn Littlewood." *Rendiconti di Palermo*, 35 (1913), 265-276.

―― 2. *Handbuch der Verteilung der Primzahlen.* Leipzig, 1909, 2 vols.

―― 3. "Über die Konvergenz einiger Klassen von unendlichen Reihen am Rande des Konvergenzgebietes." *Monatsh. f. Math. und Phys.* 18 (1907), 8-28.

K. Mahler 1. "On the Translation Properties of a Simple Class of Arithmetical Functions." *Journ. Math. Phys. M.I.T.* 6 (1927), 158-164.

BIBLIOGRAPHY

M. Plancherel 1. "Contribution à l'étude de la représentation d'une fonction arbitraire par des intégrales définies." *Rendiconti di Palermo,* 30 (1910), 289-335.

F. Riesz 1. *Comptes Rendus,* 144 (1907), 615-619 and 734-736.

—— 2. "Sur la formule d'inversion de Fourier." *Acta Szeged,* 3 (1927), 235-241.

R. Schmidt 1. "Über divergente Folgen und lineare Mittelbildungen." *Math. Zeitschr.* 22 (1925), 89-152.

E. C. Titchmarsh 1. "A Contribution to the Theory of Fourier Transforms." *Proc. Lond. Math. Soc.* (2), 23 (1924), 279-289.

C.-J. de la Vallée Poussin 1. "Sur les fonctions presque périodiques de H. Bohr." *Annales de la Société Scientifique de Bruxelles,* A 47 (1927), 141.

H. Weyl 1. "Über die Konvergenz von Reihen, die nach Orthogonalfunktionen fortschreiten." *Math. Ann.* 67 (1909), 225-245.

—— 2. "Integralgleichungen und fastperiodische Funktionen." *Math. Ann.* 97 (1926), 338-356.

N. Wiener 1. "Generalized Harmonic Analysis." *Acta Math.* 55 (1930), 117-258.

—— 2. "Tauberian Theorems." *Annals of Mathematics,* 33 (1932), 1-100.

—— 3. "Hermitian Polynomials and Fourier Analysis." *Journ. Math. Phys. M.I.T.* 8 (1929), 70-73.

—— 4. "A New Method in Tauberian Theorems." *Jour. Math. Phys. M.I.T.* 7 (1928), 161-184.

—— 5. "On the Representation of Functions by Trigonometrical Integrals." *Math. Zeitschr.* 24 (1929), 575-617.

—— 6. "The Spectrum of an Arbitrary Function." *Proc. Lond. Math. Soc.* (2), 27 (1928), 487-496.